Agrarian Science for Sustainable Resource Management in
Sub-Saharan Africa

T0316962

SUPPORT AFRICA INTERNATIONAL
Studies in sub-Saharan Africa

Vol. 3

PETER LANG
Frankfurt am Main · Berlin · Bern · Bruxelles · New York · Oxford · Wien

George Ouma
Franz-Theo Gottwald
Isabel Boergen
(eds.)

Agrarian Science for Sustainable Resource Management in Sub-Saharan Africa

PETER LANG
Internationaler Verlag der Wissenschaften

Bibliographic Information published by the Deutsche Nationalbibliothek
The Deutsche Nationalbibliothek lists this publication in the Deutsche Nationalbibliografie; detailed bibliographic data is available in the internet at <http://www.d-nb.de>.

ISSN 1614-001X
ISBN 978-3-631-58524-5

© Peter Lang GmbH
Internationaler Verlag der Wissenschaften
Frankfurt am Main 2009
All rights reserved.

Printed in Germany 1 2 3 4 5   7

www.peterlang.de

Foreword to the Series

The research findings published in the series Studies in sub-Saharan Africa have been promoted by SUPPORT AFRICA INTERNATIONAL and the Munich-based Schweisfurth Foundation.

The articles and contributions are mainly research papers submitted to the annual call for the Research Award of Sustainable Agriculture by the organisations mentioned above. There also annul award calls in ecology and inclusive education.

The findings reflect the present state of research and development projects in the respective academic fields at universities in sub-Saharan Africa. The authors are independent scientists taking responsibility for their studies published.

Publishing this series is to encourage institutional research and development at sub-Saharan universities. It appeals to members of academic departments, providing them with a platform to voice their concern. Though this series is open to a world-wide readership, it primarily aims at reaching the scientific world in sub-Saharan Africa.

Baldur Ed. Pfeiffer
SUPPORT AFRICA INTERATIONAL
University of Eastern Africa, Kenya

Franz-Theo Gottwald
Schweisfurth Fondation
Humboldt University

TABLE OF CONTENTS

Foreword to the series   5

Baldur Ed. Pfeiffer and Franz-Theo Gottwald

Foreword   9
Agrarian science for sustainable resource management in
sub-Saharan Africa
Franz-Theo Gottwald/Isabel Boergen

I. Theories on sustainable resource management

Agrarian science for sustainable resource management   12
   Anozie Edith Ngozi

Agrarian science for sustainable resource management   24
   Cyril Ifeanyi Duruigbo

Agrarian science for sustainable resource management   32
   Izuchukwu Innocent Ibeawuchi

Agrarian science for sustainable resource management   43
   G. O. Ihejirika

II. Applications of sustainable resource management

The case of indigenous resource rights in Akwa Ibom State   49
   Emmanuel O. Eyo

Fire-wood fuel practice in Benu State: Implications for land use   58
   Margaret O. Ode

Managing Africa's natural resources to mitigate land degradation: 67
A soil science perspective
   Joshua O. Ogunwole

Soil science and smallholder agriculture in sub-Saharan Africa   82
   Emmanuel Uzoma Onweremadu

Application of organic farming for sustainable horticultural   96
production to reduce poverty, improve food security and health
of rural households and environmental conservation in Kenya
   George Ouma

8

III. Special topics

Dietary ascobic acid supplementation in practical diets for African   109
Catfish Clarias gariepillus (Burchell 1822) fingerlings
   Gbadamosi, O. K., Fasakin E. A. and O. T. Adebayo

Managing strategies to foster sustainability in the poultry industry   110
in sub-Saharan Africa
   Victorian N. Meremiku

FOREWORD

AGRARIAN SCIENCE FOR SUSTAINABLE RESSOURCE MANAGEMENT IN SUB-SAHARAN AFRICA

Franz-Theo Gottwald and Isabel Boergen

The world is facing a new food crisis, due to a higher demand of animal proteins, climate change, the increased cultivation of energy crops and speculation in the raw material market. In the past decades, world agriculture has undergone severe changes. The production of raw materials and agricultural goods is industrialised, centralised, energy- and resource-intensive and far from being ecologically and socially sustainable. In the rich countries, these developments already have alarming consequences: increased food prices, crop shortages, land deprivation and negative ecological and social impacts, to mention only a few of them. But the developing countries will be impaired even more.

Land shortages, resource scarcity, population growth and environmental depletion is often said to encourage violent conflicts in developing countries. Robert Kaplan argued in his essay *The Coming Anarchy* already in 1994, that the violence and wars in Somalia, Liberia and Rwanda were highly influenced by these factors.

But the problems of resource and food shortages are not emerging from these African countries themselves – they are largely made by an unfair world trade system and the enormous demand for food and raw materials by the industrialized world. Valuable land is devastated by huge mono-cultural crop cultivations and the related excessive use of pesticides. Tropical forests, which play a crucial role in combating global climate change are slashed and burned in order to grow soy and other high-protein plants for the giant animal factories in Europe and the U.S.

Meanwhile, the local population of the developing countries suffers from hunger and malnutrition. Worldwide, 860 million people are currently starving. And 80 percent of the poorest live in Africa. On the G8 summit 2008, one main objective for future development aid was to forward rural development in Africa. And indeed, the reconstruction and development of rural communities and small-scale agriculture is urgently needed.

Small-scale agricultural production can ensure regional food security in the long term. The daily production serves the self-supply of the farmers and their families; the surplus is sold at local markets. Also, small-scale production and family businesses are sustainably managed – simply because the farmer has to look ahead and guarantee future earnings for the generations to come. Hence, there is no better way to implement generational justice in the agricultural sector than by supporting subsistence agriculture.

Also, unlike industrialized production which mainly focuses on export, small-scale agriculture is ecologically and socially compatible.

The global developments not only call for a more differentiated and regionally adapted developmental aid; they also show the need for altered political strategies in the concerning countries when it comes to land distribution, ownership regulations and subsidies.

Self-supporting farmers and their families must be kept able to sustainably manage their land. They need direct and unbureaucratic support in times of drought and food insecurity. Also, the government has to protect these families against foreign investors and landlords. This is fundamental in order to insure food security and to pacify conflict zones.

An interesting discussion paper of the Humboldt University Berlin, Department of Agricultural Economics and Social Sciences, on small-scale agro-pastoralist households in Somali, Ethiopia, showed, that resource sharing offers asset-poor households opportunities to stabilize and enhance their asset-base in drought years, providing incentives for co-operative rather than conflictive relations with intruding pastoralists. Hence, political structures and incentives are needed in order to ensure food security and peaceful coexistence in the concerning regions*.

Future objectives must not focus on short-term gains and economic profit for some few large international companies, but building up the necessary political framework and infrastructure for a nationwide food supply for the local people. It is not tolerable, that urgently needed crops are cheaply produced in the developing countries and exported in order to fatten animals in rich countries.

In order to stem these threatening developments and search for feasible alternatives, Support Africa International e. V. and the Schweisfurth-Foundation are encouraging African scientists to conduct some research which aims at a sustainable, regionally adapted agriculture. Every year, the best scientific papers are rewarded with the Research Award and published in our serial "Studies in Sub-Saharan Africa".

This year, the contest was themed "Agrarian Science for Sustainable Resource Management in sub-Saharan Africa". Therewith, we intended to highlight the great importance of science in order to face resource scarcity, and to figure out alternative and applicable models for sustainable resource management.

The studies received show, that there exist useful theoretical frameworks in agrarian and animal sciences which can help establishing and implementing practical solutions for resource management in sub-Saharan Africa. Another major focus of the studies this year was on soil science, which is tremendouslyimportant in order to combat desertification and erosion, and conserve soil fertility Furthermore, the papers offer outstanding examples of applications for different resource-sustaining methods.

References

*Bogale, Ayalneh and Benedikt Korf (2005). To share or not to share? (Non-)Violence, Scarcity and Resource Access in Somali Region, Ethiopia. ICAR Discussion Paper, Humboldt University, Berlin.

AGRARIAN SCIENCE FOR SUSTAINABLE RESOURCE MANAGEMENT

Anozie Edith Ngozi

Abstract
The productivity and intensification of agricultural systems require a broad-based participatory research and adoption of resource enhancing agricultural technologies by various stakeholders. The conservation of natural resources in the face of growing poverty, food insecurity and population growth requires an integration of agro-biodiversity, socio-economic variables and implementing environmentally friendly policies capable of boosting food security, alleviating poverty and ensuring sustainable resource management.

Introduction

Agriculture has changed dramatically, especially since the end of World War I. Food and fibre productivity soared due to new technologies, mechanization, increased chemical use, specialization, and government policies that favoured maximizing production.

Although these changes have had many positive effects and reduced many risks in farming, there have also been significant costs prominent among these are topsoil depletion, groundwater contamination, the decline of family farms, continued neglect of the living and working conditions for farm labours, increasing costs of production, and the disintegration of economic and social conditions in rural communities.

Continued agricultural growth is a necessity, not an option, for most developing countries. Further, this growth must be achieved on a sustainable basis so as not to jeopardize the underlying base of natural resources and it must be equitable if it is to contribute to the alleviation of poverty and food insecurity. Government policies, institutional development, agricultural research and projects at local, regional and national levels need to be designed and implemented with these objectives in mind. Productivity has grown fastest in irrigated agriculture, because of the increased use of modern inputs in irrigation, fertilizer, pesticides high-yielding varieties and machinery.

However, this intensification has also increased the potential for the inappropriate use of modern inputs, particularly when inappropriate incentives prevail. The following major environmental problems are associated with intensification in irrigated areas (Pingali and Rosepant, 1994), where poor drainage can lead to waterlogged soils and a rise in the water table. In arid and semi-arid areas, this in term causes salt to build up in the soil. Salinisation reduces yields and can eventually lead to abandonment of land.

- Perennial flooding of rice paddies and continuous rice culture lead to a build up of micronutrient deficiencies and soil toxicities, formation of hard pans in the soil and a reduction in the nitrogen-carrying capacity of the soil. Work at the International Rice Research Institute shows that farmers have to increase the amount of fertilizer they use over time simply to maintain existing yields in intensive paddy fields (Pingali, 1992).
- Excessive and inappropriate use of pesticides deteriorates the quality of water, poses health hazards for humans and leads to resistance of pests to pesticides. Farmers can become trapped into using more and more frequent sprays to control pest damage.
- An increasing reliance on a few carefully bred crop varieties contributes to a loss of genetic diversity and to a common vulnerability to the same pest- and weather-related risks.

In some cases, millions of hectares of land have been planted of the same wheat or rice varieties and widespread losses have occurred because of the outbreak of a single pest or disease. The loss of traditional varieties also reduces the pool of genes available for breeding plants capable of resisting evolving plant pests.

Some dramatic changes will be needed in the ways that people raise crops and livestock if much biodiversity is to survive the next 50 years. How agriculture is transformed and intensified in a sustainable manner will be the key to how many species and how much genetic variation are still around in the next century. A focus on conserving biodiversity in "protected areas" alone will not work. Protection of a sample of natural habitats is neither sufficient nor desirable to conserver biodiversity for two simple reasons:

- Most of the world's biodiversity exists in human-managed or modified systems and
- land use patterns and socio political factors in areas adjacent to parks and reserves that have major implications for the integrity biological diversity in 'protected' areas (Pimentle et al., 1992).

Given the ultimate importance of biodiversity in its broadest sense to agriculture, a strategy for main streaming biodiversity in agricultural development needs to address the off-site impacts of land use systems. Steps in this direction are outlined in the call for a new agricultural research and development paradigm. Work is already underway to addresses a range of issues related to off-site impacts, including the reduction or elimination of agricultural pollutants in groundwater, and in runoff and the placement of greater emphasis on integrated pest management (1pm) strategies. How and why rural people conserve, enhance and use biodiversity have rarely been taken into account when designing management interventions and devising policy for agricultural development and natural resource management. But the active participation of farmers, ranchers, and pastaralists - and especially resource-poor operators - is essential in the design and implementation of biodiversity and agricultural development projects. (Thrupp, Cabarle, and Zazueta, 1994; Wilcox and Duin, 1995).

The evolving agricultural research paradigm includes, but is not restricted to, the adopting integrated pest management. 1pm strategies includes the release of bio-control agents, deployment of genetically resistant cultivars and breeds, more judicious use of pesticides and herbicides, alteration of cropping patterns to thwart the building of pests and disease, and placement of greater emphasis on crop rotation, where economically feasible to retard soil degradation and reduce pest pressure.

The use of participatory approach with farmers, two types of on-farm research are typically found: demonstration plots on farmers' land and experimental work that involves farmers and other stakeholders in the design of models from the "ground up". Much more of the second type of on-farm research is needed that involves farmers, pastoralists and other "clients" of agricultural research and development from the inception of the study design. In this manner, research would be more demand driven. Improve the use of indigenous knowledge. How and why local people use natural resources can provide important information for more appropriate agricultural research and development efforts.

Support research, development, and dissemination of lesser-known crops and animal neglected traditional varieties and breeds, many of which are particularly well suited to difficult environments, would be included in this broad research effort. But sustained support for research on the major food and industrial crops as well as livestock remains essential. Support research on new crops and livestock scope exists for new crops and livestock to fill specially market and environmental niches. In some cases, natural vegetation communities could be managed for the production of new domesticated animals. A deeper commitment to research on crop and livestock candidates would thus underscore the value of conserving biodiversity and natural habitats.

Biodiversity in managed landscapes often best served by promoting a mixture of land uses that provides varied habitats for wildlife adapted to altered areas. Achieved a greater diversity of habitats within land use systems, biodiversity within a land use system, such as intensive cereal cropping, can be achieved by allowing for a variety of habitats, such as riparian butter strips, shelter belts, windbreaks, strip cropping and wetlands. Diversity of habitats on the landscape creates more niches for wildlife, some of which are beneficial in controlling crop pest. More diverse habitats, including managed ones, also promote the more efficient use of nutrients and create microclimates that butter crops from inclement weather.

Recycle organic matter through measures such as incorporating livestock or grain manure-no-till or minimum-till farming, help to sustain the diversity of soil micro-organisms, which are so important in nutrient recycling.

Focus research on lifetime and herd productivity characteristics. Deterministic simulation models and live animal experimentation can be used in some cases to achieve these goals. Determine the critical number of breeds for conservation purposes. Analysis of genetic "spacing" between breeds and to identify those breeds that are significantly different or unique from others. Learn more about genetic components of adaptation in livestock. A better understand of traits such as resistance to ticks and use of body reserves would aid breeding efforts and would likely underscore the importance of safeguarding so called "minor" breeds.

Many national agricultural research and extension systems have yet to integrate environmental concerns successfully into their agenda. Too little consideration is given to the sustainability features of recommended technologies, to broader aspects of natural resource management problems of more fragile rain-fed areas where resource degradation is considerable.

This research paper is aimed at addressing the issues of poor resource management through sustainable agricultural practices and proper integration of environmentally-friendly technologies needed to boost food production, alleviate poverty and conserve natural resources.

Managing natural resources sustainable agriculture

Sustainable agriculture integrates three main goals: environmental health, economic profitability and social economic equity. A variety of philosophies, policies and practices have contributed to these goals. People in many different capacities, from farmers to consumers have shared this vision and contributed to it. Despite the diversity of people and perspectives, the following themes commonly weave through definitions of sustainable agriculture.

Sustainability rests on the principle that we must meet the needs of the present without compromising the ability of future generations to meet their own needs. Therefore, stewardship of both natural and human resources is of prime importance. Stewardship of human resources includes consideration of social responsibilities such as working and living conditions of labourers, the needs of rural communities and consumer health and safety both in the present and the future. Stewardship of land and natural resources involves maintaining or enhancing this vital resource base for long term.

A systems perspective is essential to understanding sustainability. The system is envisioned in its broadest sense, from the individual farm to the local ecosystem and to communities affected by this farming system both locally and globally. An emphasis on the system allows a larger and more thorough view of the consequences of farming practices on both human communities and the environment. A system approach gives us the tools to explore the interconnections between farming and others aspects of our environment.

A system approach also implies interdisciplinary efforts in research and education. This requires not only the input of researchers from various disciplines, but also farmers, farm workers, consumers, policymakers, and others. Making the transition to sustainable agriculture is a process. For farmers, the transition to sustainable agriculture normally requires a series of small realistic steps.

Finally, it is important to point to the fact that reaching the goal of sustainable agriculture is the responsibility of all participants in the system, including farmers, labourers, policymakers, researchers, retailers, and consumers. Each group has its own part to play, its own unique contribution to make to strengthen the sustainable agriculture community. When the production of food and fibre degrades the natural resource base, the ability of future generations to produce and flourish decreases. The decline of ancient civilizations in Mesopotamia, the Mediterranean region, pre-Columbian Southwest U.S., and Central America is believed to have been strongly influenced by natural resources degradation from non-sustainable farming and forestry practices. Water is the principal resource that has helped agriculture and society to prosper, and it has been a major limiting factor when mismanaged.

Sustainable production practices involve a variety of approaches. Specific strategies must take into account topography, soil characteristics, climate, pests, local availability of inputs and the individual grower's goals. Despite the site-specific and individual nature of sustainable agriculture, several general principles can be applied to help growers select appropriate management practices.

Properly managed, diversity can also butter a farm in a biological sense. For example, in annual cropping systems, crop rotation can be used to suppress weeds pathogens and insect pests. Also cover crops can have stabilizing effects on the agro-ecosystem by holding soil and nutrients in place, conserving soil moistures with moved or standing dead mulches and increasing the water infiltration rate and soil water holding capacity. Cover crops in orchards and vineyards can butter the system against pest infestations by increasing beneficial arthropod populations and can therefore reduce the need for chemical inputs. Using a variety of cover crops is also important in order to protect against the failure of a particular species to grow and to attract and sustain a wide range of beneficial arthropods. Optimum diversity may be obtained by integrating both crops and livestock in the same technology and government policy.

Common philosophy among sustainable agriculture practitioners is that a 'health' soil is a key component of sustainability, that is, healthy soil will produce healthy crop plants that have optimum vigour and are less susceptible to pests. While many crops have key pest that attack even the healthiest of plants, proper soil, water or nutrient imbalance. Furthermore, crop management systems that impair soil quality often result in greater inputs of water, nutrients, pesticides and/or energy for tillage to maintain yields

In sustainable systems, the soil is viewed as a fragile and living medium that must be protected and nurtured to ensure its long-term productivity and stability. Methods to protect and enhance the productivity of the soil include using cover crops, compost and/or manures, reducing tillage, avoiding traffic on wet soils and maintaining soil cover with plants and or matches. Conditions in most California soils (warm, irrigated and tilled) do not favour the build-up of organic matter. Regular additions of organic matter or the use of cover crops can increase soil aggregate stability, soil tilth and diversity of soil microbial life.

Many inputs and practices used by conventional farmers are also used in sustainable agriculture. Sustainable farmer, however, maximize reliance on nature renewable and on-farm inputs. Equally important are the environmental, social and economic impacts of a particular strategy. Converting to sustainable practices does not mean simple input substitution. Frequently, it substitutes enhanced management and scientific knowledge for conventional inputs, especially chemical inputs that harm the environment on farms and in rural communities. The goal is to developed efficient biological systems which do not need high levels of material inputs.

Thorough integration of natural processes such as nutrient cycling, nitrogen fixation and pest-predator relationships into agricultural production processes ensure profitable and efficient food production. Minimization of the use of those external and non-renewable inputs, with the potential to damage the environment or harm the health of farmers and consumers and targeted used of the remaining inputs used, minimize the costs.

The full participation of farmers and other rural people in all processes of problem analysis and technology development, adaptation and extension lead to an increase in self-reliance among farmers and rural communities, so a greater productive use of local knowledge and practises, including innovative approaches not yet fully understood by scientists or widely adopted by farmers, and also the enhancement of wildlife and other public goods of country side. Sustainable agriculture seeks the integrated use of a wide range of pest, nutrient, soil and water management technologies. It aims for an increased diversity of enterprise within farms combined with increased linkages and flows between them. By-products or wastes from one component or enterprise become inputs to another. As natural processes increasingly substitute for external inputs the negative impacts on the environment are reduced and positive contributions are made to regenerate natural resources.

Increasing evidence shows that regenerative and resources-conserving technologies and practices can bring both environmental and economic benefits for farmers, communities and nations. The best evidence comes from countries of Africa, Asia and Latin America, where the concern is to increase food production in the areas where farming has been largely untouched by the modern packages of externally supplied technologies. In these lands, farming communities adopting regenerative technologies have substantially improved agricultural yields, often using only a few or no external inputs. (Bunch, 1990, 1993; GTZ, 1992; UNDP, 1992; Krishna, 1994; Shah, 1994; SWCB, 1994; Balbarin and Alcober, 1994; de Freitas, 1994 and Pretty, 1995b).

The high-inputs and generally irrigated lands, farmers adopting regenerative technologies have maintained or improved yields while substantially reducing their use of inputs (Bagadion and Korten, 1991). The basic challenge for sustainable agriculture is to make better use of available physical and human resources. This can be done by minimizing the use of external inputs, by regenerating internal resources more effectively, or by combining two in various ways. This ensures the efficient and effective use of what is available and keeps any dependencies on external systems to a reasonable minimum.

Implementing sustainable agrarian policies that reduce resources degradation Policy interventions that seek to overcome environmental problems in agriculture need to be based on a proper understanding of why farmers degrade natural resources. For example, farmers often seem to overgraze rangeland depleting soil nutrients and organic matter and overuse irrigation water, pesticides and nitrogen, whereas these actions cause health problems and reduce future incomes for themselves, their children and the communities in which they live.

The answer lies with incentives including discount rates. Farmers are not irrational. To the contrary, they maximize income and minimize risks in a dynamic context and often wider harsh conditions and serious constraints. For example, they degrade resources when there are good economic and social reasons for doing so. If the management of natural resources is to be improved, these economic and social incentives will need to be changed in appropriate ways.

Current policies often do not reflect the long-term social and environmental costs of resource use. The external costs of modern farming, such as soil erosion, health damage, or polluted ecosystems, generally are not incorporated into individual decision making by farmers. In this way, resource-degrading farmers bear neither the costs of damage to the environment or economy, nor those incurred in controlling the polluting or damaging activity (Pretty, 1996).

In principle, it is possible to imagine pricing the free input to farming of the clean, unpolluted environment. If charges were levied, in some way, then degraders or polluters would have higher costs, would be forced to pass them on to consumers, and would forced to stitch to more resource-conserving technologies. This notion is captured in the polluters-pays principle (OECD, 1989).

However, beyond the notion of encouraging some internalisation of costs, it has not been of practical use for policy formulation in agriculture. Although a growing number of policy initiatives are oriented specifically toward improving the sustainability of agriculture, most have focused on input reduction strategies. Only a few as yet represent coherent plans and processes that clearly demonstrate the value of integrating policy goals. A thriving and sustainable agricultural sector requires both integrated action by farmers and communities and integrated action by policy makers and planners. This implies both horizontal integration with better linkages between sectors and micro and macro level.

Integration has been the policy buzz of the 1990s. But putting this desired integration into practices has been much more difficult. There have been substantial differences in the views of major policy actors, such as those representing the interests of farmers, environmentalists and treasuries.

In addition to strategies for preserving natural resources and changing production practices, sustainable agriculture requires a commitment to changing public policies, economic institutions and social values. Strategies for change must take into account the complex, reciprocal and ever-changing relationship between agricultural production and the broader society.

The "food systems" extends far beyond the farm and involves the interaction of individuals and institutions with contrasting and often competing goals including farmers, researchers, input suppliers, farm workers, unions, farm advisors, processors, retailers, consumers, and policymakers. Relationships among these actors shift overtime as new technologies spawn economic, social and political changes.

Governments can do much with existing resources to encourage and nurture the transition from modernized systems toward more sustainable alternatives, declaring a national policy for sustainable agriculture. This would help to raise the profile of these processes and needs as well as give explicit value.

Finally, for sustainable agricultural to spread widely policy formulation must be enabling and create the condition for sustainable development based on locally available resources, local skills and knowledge.

Managing natural resources through participatory research

Agricultural research has greatly increased the yields of important staple food crops and livestock products, yet many people have meant more food availability and trade opportunities. But some people in rural areas in developing countries still live in abject poverty. Therefore, policy makes, donors and researchers are refocusing their priorities away from simply producing more food to make sure that agricultural research benefits the poor in particular.

How agriculture can be intensified without damaging biodiversity is a critical question for rural development. Environmentally inappropriate intensification of agriculture has led to eutrophication of lakes and estuaries, loss of soil micro-organisms, accelerated soil erosion, contamination of groundwater, and draining of wetlands. All of these activities trigger a potentially dangerous loss of biodiversity. But wild species are essential for agricultural improvement, because they are the source of new economic plants and animals and provide important services such as pollination and pest control.

Possible remedial measures can be adopted to address the loss of biodiversity associated with agricultural development such as minimizing habitat fragmentation by providing wildlife corridors along "bridges" of natural habitat. By shifting to integrated pest management (1pm) strategies, such as rotating crops and relying on bio-control agents to check crop and livestock pest and eliminating fiscal and regulatory measures that promote homogeneity in crop and livestock production.

The new vision for agricultural research adopts a holistic approach that is more sensitive to environmental concerns, while still addressing the need to boost the yields and incomes of rural producers and caretakers of the land. This includes, but is not restricted to:

- Integrated pest management
- A participatory approach with farmers
- Better use of farmers knowledge
- Greater support for research, development and dissemination of lesser-known crops and animals.

- Support for research on new crops and livestock.
- Greater sensitivity to the value of a mosaic of land uses
- Greater diversity of habitats with land use systems
- Greater reliance or recycling of organic matter
- The shift of the research focus from individual traits to lifetime and herd productivity characteristics.
- Determination of the critical number of breeds for conservation purposes and an effort to learn more about the genetic components of adaptation in livestock.

This notion of a new research paradigm has implications for institutional development and the exploration of new ways of doing business. Innovative institutional arrangements would include more effective partnerships among agricultural research centres, NGOs, growers, associations, private companies involved in the manufacture and sale of agricultural technologies, universities and agricultural extension agencies, and development lending institutions. To some degree, all of these partnerships are being explored and tested.

Generating the gains in agricultural productivity necessary to secure food availability and livelihoods in the developing world over the coming decades requires an approach in which the intensification of agricultural systems is consistent with the conservation of natural resource base. This approach requires less reliance on the intensive use of external inputs and greater dependence on management skills and location-specific knowledge of agro-ecosystems. Integrated pest management constitutes one of such an approach and is critical to sustainable rural development. (Pingali and Gerpacio, 1997).

The research processes should begin with the careful identification of socio-economic and environmental problems affecting the agricultural productivity, followed by a research strategy aimed at improving the genetic resources, crop/livestock farming systems as well as natural resource management. The research priorities should be based on agro-ecological, biophysical and demographic parameters so as to ensure a sound and sustainable research result.

Agricultural research can help to alleviate poverty in many ways farm household that adopt resulting technologies can benefit from higher yields and incomes, but benefits are not just felt by the adopting households. The indirect impacts of research (such as cheaper food and more jobs) can also improve the living standards of wider populations. Impact can also be negative. All are important, and all should be included in assessment of impacts. Technology should be tailored to fit people's livelihood strategies, and it should be targeted at areas where agriculture still plays a significant role in the lives of poor farmers.

Finally, to increase the impact of agricultural research on poverty, research organizations need to embrace a culture of institutional learning and change (ILAC). This can be fostered by a spirit of critical self-awareness among professionals and an open culture of reflective learning within organizations.

Conclusion
The sustainability of natural resources in the developing countries could be achieved through implementing sound resource-building policies and by a holistic participatory research approach built on agro-ecological, socio-economic, demographic and biophysical framework.

References

Bagadion, B. U. and F. F. Korten, 1991. "Developing Irrigators' Organizations: A Learning Process Approach", In: Michael M. Cernea (ed.), Putting People First, 2d ed, New York, Oxford University Press.

Bunch, Roland, 1990. Low-input Soil Restoration in Honduras: The Cantarranas Farmers to Farmer Extern Program. Gatekeeper series, SA 23. London: International Institute for Environment and Development, Sustainable Agriculture Program. Institutional Learning and Change in the CGIAR. Impact Assessment Discussion Paper, (Summary Record of the Workshop Held at IFPRI, Washington, D.C, February 2003.

OECD (Organisation for Economic Cooperation and Development), 1989. Agricultural and Environmental Policies, Paris.

Pingali, Prabhu L., 1992. "Diversifying Asian Rice Farming Systems: A Deterministic Paradigm", In: Shawki Barghouti, L. Garbux, and Dinaumali, (eds.), Trends in Agricultural Diversification: Regional Perspectives, pp 107-26. World Bank Technical Paper 180, World Bank, Washington. D.C.

Pingali and Rosepant, 1994. Confronting the Environmental Consequences of the Green Revolution in Asia. Environment and Production, Technology Division. Discussion Paper 2. Washington, D.C: International Food Policy Research Institute.

Pigali, Prabhu L. and R. V. Gerpacio, 1997. "Living with Reduced Insecticide Use in Tropical Rice". Food Policy 22(2): 107-18.

Pimentle, D. U., Stachow, D. A., Takaces, H. W., Brubakar, A. R., Dumes J. J., Mency, J. A., O'Neil S., Onsi, D. E. and D. B. Corzilius, 1992. Conserving Biological Diversity in Agricultural Forestry System. Bioscience 42(5): 354-62.

Thrupp, L. A., Cabarle B. and A. Zazueta, 1994. "Participatory Methods in Planning and Political Processes: Linking the Grassroots and Policies for

Sustainable Development," In: Agriculture and Human Valves, 11 (2-3): 77-84.

Wilcox, B. A. and K. N. Duin, 1995. "Indigenous Cultural and Biological Diversity: Overlapping Valves of Latin American Eco-regions," In: Cultural Survival Quarterly, 18 (Winter): 49-53.

AGRARIAN SCIENCE FOR SUSTAINABLE RESOURCE MANAGEMENT

Cyril Ifeanyi Duruigbo

Abstract
The degradation of natural resources is a global problem that threatens the livelihood of millions of poor people in Africa, Asia, Middle East and Latin America occasioned by misuse of modern inputs, poverty, unequal access to resources, inappropriate policies and population growth. The adoption of sustainable agricultural practices and agro-ecological approach to farming as well as implementing policies that promote biodiversity conservation in agriculture must be embraced by all stakeholders to boost food security, alleviate poverty, and ensure sustainable resource management.

Introduction

Agriculture is an important production sector of the economy in many developing countries as well as the principal means of livelihood for most resource-poor rural communities in the world. Farmers are the custodians of a nation's natural resources, and have a direct impact or bearing on how these resources are managed. Thus, agricultural development should simultaneously contribute to four principal goals: namely growth, poverty reduction, food security and sustainable natural resources management. The new priority for environmental sustainability that has emerged in the 1990s does not negate the need for agriculture to continue contributing to growth, poverty alleviation and increased food security rather it is of the opinion that agriculture is required to do this in ways that do not degrade natural resources of the environment. The degradation of natural resources is a global problem that threatens the livelihood of millions of poor people in Africa, Asia, Middle East and Latin America. Two fundamentally different types of environmental problems are associated with agriculture. Firstly, the misuse of modern inputs (irrigation, water, fertilizers and pesticides) in intensive farming systems. This misuse is related to poor management of inputs, inappropriate policies that encourage farmers to overuse inputs, and externality problems that lead farmers to undervalue environmental costs and benefits.

The second type of environmental problems arises from extensive farming systems with an associated rapid population growth, poverty and growth rates in agricultural productivity that is insufficient to combat food insecurity. Farmers in these areas typically use low levels of modern inputs, reduce fallows and expand cultivation into marginal and environmentally fragile areas. Poverty and insufficient agricultural intensification are the fundamental problems, but resource degradation is often worsened by insecure property rights, externality problems, high costs of collective action for managing properties or undertaking resource-improving investments and inappropriate government policies. The environment must be managed in a sustainable manner to avoid degradation by ensuring its long term capacity to provide the goods (natural resources) and services (eco-systems functions) in which human and agricultural development depends, the need to secure equitable access by the poor to environmental assets such as food and health in particular and by reducing their vulnerability to environmental related risks in general.

The conservation of natural resources often requires collective action by groups of users or stakeholders. Organizing farmers into effective and stable groups for collective action is difficult and success is conditioned on a range of physical, social and institutional factors (Uphoff, 1986; Ostrom, 1994; Rasmussen and Meinzen-Dick, 1995).

Innovation by research centres, development organizations and farmers themselves have produced many promising technologies and practices, for making agriculture and natural resource management more sustainable. Promoting sustainable natural resource management requires an understanding of the interaction between local and external institutions which must build on local strengths (Ruth-Meingen-Dick, 1995). A reversal of environmental degradation requires a new livelihood options that change peoples incentives such as the benefits and costs of resource use. When innovation in resource management is driven by perceived tradeoffs, participatory assessments of livelihood strategies are important for developing a common understanding of how these depend on natural resource assets (Carney, 1998). The global concern for the depletion of natural capital stocks is not only an expression of the conservation ethics, but is linked to concern with international poverty, famine and disaster. Ecological threats of global significance are paralleled by the vulnerability of over 800 million poor people to malnutrition, disease, high rates of infant mortality coupled with rising inequality in the distribution of wealth. The capacity of poor households, communities and countries to recover from external shocks such as war, famine, epidemic disease, hurricanes, global climate change and indebtedness partly depends on the status of their stocks of natural capital.

A new agricultural research paradigm must encourage the use of bio-technology and transgenic crops in boosting food production and sustaining natural resources. Thus the adoption of integrated pest management through using bio-control agents and the release of genetically resistant crop cultivars and breeds will reduce the use of pesticides and herbicides that pollute the environment and threaten biodiversity. Given the ultimate importance of biodiversity in agriculture, a strategy for mainstreaming biodiversity in agricultural development needs to address the off-site impacts of land use systems. This research paper will explore the role of agro-ecology, re-generative agricultural practices and the need to implement sustainable resource improving policies in fostering food security, poverty alleviation and sustainable resource management.

Sustaining natural resource management through
Agro-ecology and regenerative agricultural practices

Sustainable agriculture as a practice and research priority has emerged in response to widespread recognition of the need to balance food production with environmental and social health (National Research Council, 1989; Pretty, 1995). The sustainability of increasing mechanized, high-input and specialized approaches is being questioned, due to its negative impact on the environment.

The United States Congress (1996) defined sustainable agriculture as "An integrated system of plant and animal production practices having a site-specific application that will over the long run: (a) satisfy human food and fibre needs, (b) enhance environmental quality and the natural resource base upon which agriculture economy depends, (c) make the most efficient use of non-renewable resources and on-farm resources and integrate where appropriate natural biological cycles and controls, (d) sustain the economic viability of farm operations and (e) enhance the quality of life for farmers and society as a whole."

Agro-ecology which involves the use of ecological principles for the design and management of sustainable and resource conserving agricultural systems offers several advantages over the conventional agronomic or agro-industrial approach. This is because agro-ecology relies on farmers' indigenous farming knowledge and selected modern technologies to manage diversity and incorporate biological principles and resources into farming systems to boost food production.

The concept of sustainable agriculture seeks the integrated use of a wide range of pest, nutrient, soil and water management technologies aimed at increasing diversity of enterprises within farms, combined with increased linkages and flows between them. Bye-products or wastes from one component or enterprise become inputs to another. As natural processes increasingly substitute for external inputs, the negative impacts on the environment are reduced and positive contributions are made to regenerate natural resources.

Increasing evidence shows that regenerative and resource conserving technologies and practices can bring both environmental and economic benefits to farmers, communities and nations. The best evidence comes from countries of Africa, Asia and Latin America, where the concern is to increase food production in the areas where farming has been largely untouched by the modern packages of externally supplied high input technologies. In these areas, farming communities adopting regenerative technologies have substantially improved agricultural yields often using only a few or no external inputs (Bunch, 1990; UNDP, 1992; Shah, 1994; Pretty, 1995). Also in the high-input and generally irrigated lands, farmers adopting regenerative technologies have maintained or improved yields while substantially reducing their use of external inputs (Bagadion and Korten, 1991; Kenmore, 1991; UNDP, 1992). In industrialized countries farmers have been able to maintain profitability even though high-input use has been cut dramatically such as in the Unites States (Liebhardt et al., 1989; National Research Council, 1989; Hewitt and Smith, 1995), and in Europe (El Titi and Landes, 1990; Vereijken, 1990; Somers, 1997).

The International Institute for Environmental and Development has examined the extent and impact of sustainable agriculture in selected countries and observed that both the governmental and nongovernmental programme included in their analysis have the following characteristics: They make use of resource conserving technologies in conjunction with group or collective approaches to agricultural improvement and natural resource management. They put participatory approaches and farmer centred activities at the centre of their agenda, so that these activities are occurring on local people's terms and are more likely to persist at the end of the project. They avoided the use of subsidies to attract local people in terms of adopting the low input technologies so that improvements will not fade away at the end of the project. By involving the active participation of women as key producers and quality facilitators and by emphasizing quality to agricultural products through agro-processing, marketing and other off-farm activities thus create employment, income and surplus retention in the rural economy.

Poverty, low agricultural productivity and natural resource degradation are severe inter-related problems in less favoured areas of the tropics. Less-favoured areas include lands that have low agricultural potential due to rainfall uncertainty, poor soils, erosion areas or other bio-physical constraints as well as areas that may have high agricultural potential but have limited access to infrastructure, markets, low population density or other socio-economic constraints.

According to a recent study by the Technical Advisory Committee of the Consultative Group on International Agricultural Research, nearly two-thirds of the rural population of developing countries or almost 1.8 billion people live in less favoured areas which include: semi-arid and arid tropics of Africa, South Asia, South America, Central America, Southeast Asia, large portions of the humid tropics of Africa and Latin America.

Strategies for developing and disseminating technologies must take into account the special characteristics and demands of less favoured areas. A high degree of diversity in biophysical and socio-economic conditions is one of the main challenges. Other challenges may include susceptibility to drought, pests, diseases, temperature extremes, fragile lands and the subsistence orientation of farmers in these areas. A technological strategy should therefore be participatory and demand-driven, stimulating and building upon adoption to local circumstances. Technologies that help reduce risks by increasing tolerance to drought, pests, diseases, or foist and conserve resource may be more effective than those that simply promote high yields in response to high levels of inputs.

Sustainable and profitable technologies are needed to conserve and efficiently use scarce water, control erosion, restore soil fertility and increase the supply of useful biomass. Such technologies are often labour or land intensive which may be unattractive to farmers where labour costs are high or where land is scarce. Labour or land-saving technologies such as improved fallows during a short rainy season or agro-forestry on farm boundaries may have more potential. In areas with limited rainfall, scarcity of biomass and high demands for alternative use of biomass (for fodder and fuel) limit the potential of many organic approaches to land management. In such circumstances technologies and policies for conserving water and profitably increasing the production of useful biomass such as promotion of woodlots should have high priority. Strategies for less-favoured areas will be most effective if they are linked to the development pathway with comparative advantage in particular circumstances. Small-scale irrigation development is likely to yield the highest returns in areas with good market access and otherwise suitable soil condition, since this can enable high-value crop production as well as intensified food crop production.

The research strategy must ensure effective risk management. Risks of crop failure due to bad weather or pests can discourage investments by farmers in land improvements and their adoption of higher-yielding technologies. Such agricultural research effort should help reduce risk by improving drought resistance in crops or developing better ways of conserving soil moisture.

Sustainable agricultural practices that make best use of natural resources and services, integrating agro-ecological processes into food production as well as reducing negative side effects on the environment and health offer a broad-based strategy approach towards sustainable resource management.

Implementing policies that encourage sustainable agriculture and reduce resource degradation

Current policies by some governments do not reflect the long term social and environmental costs of resource use. The external costs of modern farming such as soil erosion, health damage or polluted *ecosystems* generally are not incorporated into individual decision making by farmers. In this way resource-degrading farmers bear neither the costs of damage to the environment or economy or those incurred in controlling the polluting or damaging activity (Pretty, 1996). Although a growing number of policy initiatives are oriented specifically towards improving the sustainability of agriculture, most have focused on input reduction strategies. There is need to integrate sustainable agricultural policies into the farming systems. This is because a thriving and sustainable agricultural sector requires both integrated action by farmers and communities and integrated action by policy-makers and planners. This implies both horizontal integration with better linkages between sectors and vertical integration with better linkages between the micro and macro levels. Governments can do much with existing resources to encourage and nurture the transition from modernized systems towards more sustainable alternatives. Policies needed to encourage sustainable agriculture must create the enabling environment for locally generated and adopted technologies to flourish, rather than prescribing the practices that farmers should use. Such policies must be locally oriented compatible and based on locally available resources skills and knowledge. Effective government policy will have to recognize that policy is the net result of actions of different interest groups pulling in complementary and opposing directions and not just the normative expression of governments and thus seek to bring together a range of actors and institutions for creative interactions, joint learning and participation. Major changes must be made in policies, institutions, research and development to ensure the adoption of agro-ecological alternatives to farming. Policies that encourage subsidies on use of high inputs (chemicals, fertilizers, etc.) must be removed; also institutional structures, partnerships and educational process must be changed through government policy to enable the agro-ecological model grow and develop.

The challenge is thus to encourage the implementation of government policies that increase investment and research in agro-ecology and scale up projects that have proven successful to create a sound impact on poverty alleviation, food security and sustainable resource management. Government should enact laws stipulating penalties to offenders found engaging in resource-degrading agricultural practices to ensure strict compliance to environmentally-friendly practices needed to boost food production and conserve natural resources.

Conclusion

The global trend of natural resource degradation due to inappropriate policies, misuse of production inputs, social inequities, poor access to use of resources and population growth is on the increase in the less developed countries. A new agrarian science research policy that will ensure greater participation in sustainable agricultural practices, application of agro-ecological and agro-biodiversity principles in farming is urgently needed to transform and improve food security, alleviate poverty and ensure sustainable resource management.

References

Bagadion, B. U. and F. F. Korten, 1991. "Developing irrigators" Organizations. "A Learning Process Approach", In: Michael M. Cornea (ed.), Putting People First. 2nd ed. New York, Oxford University Press.

Bunch, Roland, 1990. Low-Input Soil Restoration in Honduras: The Cantarranas Farmer-to-farmer Extension Programme Gatekeepers Series. SA 23, London. International Institute for Environment and Development. Sustainable Agriculture Programme.

Carney, D., 1998. Sustainable Rural Livelihood: What Contributions Can We Make? DFID London.

Conway, G., 1985. "Agro-ecosystems Analysis", In: Agriculture Administration, 20:31-55.

Conway, G., 1997. The Doubly Green Revolution: Food for All in the 21st Century, Penguin, London.

Daily, G. C., 1997. "Management Objectives for the Protection of Ecosystem Services", Environmental Science and Policy, 3 (6) 333-339.

EI Titi, Adel and H. Landes, 1990. "Integrated Farming System of Lautenbach: A practical contribution towards sustainable agriculture", In: C. A. Edwards, R. Lai, P. Madden, R. H. Miller, and G. House (eds.) Sustainable Agricultural Systems. Ankeny IOWA. Soil and Water Conservation Society.

Hatch, J., 1976. "The corn farmers of Motupe". A Study of Traditional Farming Practices in Northern Coastal Peru. Land Tenure Centre Monograph. University of Wisconsin, Madison.

Hewitt, Tracy I. and Katherine R. Smith, 1995. "Intensive Agriculture and Environmental Quality", Examining the Newest Agricultural Myth, Greenbelt, MD, Henry Wallace Institute for Alternative Agriculture.

Kenmore, Peter, 1991. How Rice Farmers Clean up the Environment, Conserve Biodiversity, Raise More Food, Make Higher Profits. Indonesia IPM - A Model for Asia. Manila FAO.

Liebhardt, W., Andrews, R. W., Culik, M. N., Harwood, R. R. Janke, R. R., Radke J. K. and S. L. Riegger-Sachwartz, 1989. "Crop production during conversion from conventional to low input methods", In: Agronomy Journal 81 (2) 150-159.

National Research Council, 1989. Alternative Agriculture. Washington D. C. Natinal Academy Press.

Ostrom, Elinor, 1994. Neither Market nor State: Governance of Common Pool Resources in the 21st Century. Lecture Series 2, Washington D. C. IFPRI.

Pearce, D., Barlbier, E. and A. Markandya, 1990. Sustainable Development Economics, Environment in the Third World. Edward Elgar Publishing, Aldershot, U. K.

Pretty, Jules N., 1995. Regenerating Agriculture: Policies and Practices for Sustainability and Self-reliance. London, Earthscan Publications, Ltd.

Pretty Jules N. and John Thompson, 1996. "Sustainable Agriculture at the Overseas Development Administration" (ODA). Report for National Resources Policy. Advisory Department (ODA), London.

Rasmussen, L. N. and Ruth Meinzen-Dick, 1995. Local Organisations for National Resource Management: Lessons from Theoretical and Empirical Literature. Environment and Production Technology Division. Discussion Paper 2, Washington D. C. IFPRI

Shah, Parmesh, 1994. Village Managed Extension Systems in India. "Implications for Policy and Practice". In: Jan Scoones and John Thompson (eds). Beyond Farmer First. London, Intermediate Technology Publications.

Somers, B. M., 1997. Learning About Sustainable Agriculture: The Case of Dutch Arable Farmers". Niels Roling and W. A. E. Wagemakers (eds). Social Learning for Sustainable Agriculture, London, Cambridge University Press.

UNDP (United Nations Development Programme,1992). The Benefits of Diversity. An Incentive towards Sustainable Agriculture, New York.

Uphoff, Norman, 1986. Local Institutional Development: An Analytical Source Book with Cases. West Hartford. Kumarian Press.

Vereijken, Peter, 1990. "Research on Integrated Arable Farming and Organic Mixed Farming in the Netherlands". C. A. Edwards, R. Lai, R. Madden, R. H. Miller and G. House (eds.), Sustainable Agricultural Systems. Ankeny IOWA. Soil and Water Conservation Society.

AGRARIAN SCIENCE FOR SUSTAINABLE RESOURCE MANAGEMENT

Izuchukwu Innocent Ibeawuchi

Abstract
Poverty and hunger is a twin sister, which pervades much of the world's rural communities. Agrarian science through research and extension linkages had developed series of agricultural packages which through well developed manpower (educated farmers and other resource personnel) in agriculture can help to achieve sustainability in agricultural resource management to increase food production for the teeming world population thereby reduce poverty and hunger.
Keywords*: Agrarian sciences, sustainable resources management, poverty and hunger reduction.*

Introduction
The place of agrarian science in the evolution of human kind cannot be over emphasized. In the face of dwindling food for the family, farmers search for a sustainable strategy for effective and better resource management through research in agriculture, which could lead to significant improvement in their socio-economic conditions. This was in the struggle for survival in the face of hunger and poverty, which pervades much of the world's rural communities. This situation caused by inadequate resources such as inputs (fertilizers, improved seeds and cuttings, pesticides, etc.) allocation and management with its attendant problems make the entire system unsustainable. Healthy population of the rural Africa, Asia and Latin America were forced to look for the means of livelihood outside the agricultural domain, thereby undergoing deagrarianisation and more specifically depeasantisation (ODI, 2000). Deagrarianisation is defined as a long-term process of occupational adjustment, income earning, reorientation, social identification and spatial relocation of rural dwellers away from strictly agricultural based modes of livelihood (Bryceson et al., 2000). On the other hand, depeasantisation represents a specific form of deagraranisation in which peasantries lose their economic capacity and social coherence and demographically shrink in size. Perceiving these ugly trends, the UN General Assembly in the September 2000 agreed to a set of Millennium goals, which centred on halving extreme poverty and hunger, achieving universal education, halting the spread of AIDS and other diseases and reversing the loss of environmental resources (NEPAD, 2002). These goals were drawn to revitalize

and reinvigorate the rural communities to uplift their spirit to have hope for a better future.

Wars, HIV/AIDS, hunger and poverty, environmental degradation, loss of diversity, etc., are some of the teething problems giving the rural farmers sleepless nights. All these can be brought to bear with the realities on the ground through well-articulated and developed agrarian policy that can sustain the teeming population by making food available and affordable.

The understanding of this scenario gives an insight into what agriculture can achieve. Generally, agriculture is the backbone of many strong world economies and this will continue to be the main source of employment accounting for more than 40% of world population, thus making agriculture by far the most common occupation (FAOSTAT, 2002 estimate). It, however, accounts for only 44% of the world gross product on aggregate at all Gross Domestic Product (GDP) (FAOSTAT, 2005 estimate).

Agrarian societies and agriculture are intertwined and dynamic from all perspective. Changes in inputs distribution, technology and food go hand in hand with changes in power for policy formulation, wealth for employment generation including goods and services provision and participation in good governance and justice. These are interdependent for stable economic growth and sound sustainable resource management, which is centred on agrarian policy initiatives.

Food production is centred on the farmer and his use and management of input resources and other variables. The researchers and farmers are at the centre of crop improvement and resource management for sustainable food production so that the future can be assured. The management of the physical biological, energy and human/institutional resources can create wealth and food for the ever increasing world population. Agriculture provides food for man raw materials for industry, employment for the people, foreign exchange for government, etc.

This paper deals with agrarian science for sustainable resource management, appraising critically the resources available and the role of government and packages provided through research and extension in agriculture to build a sustainable environment for increase food production to reduce hunger and poverty.

The major resources for agricultural production and poverty reduction are:

Human Resources

There is much that farmers do not have control over and what they do control, they control through people. How these people are hired managed and motivated makes a huge difference. Labour management is much more than forms and paper work. It is more about finding creative new ways of increasing production and reducing loss (Anon, 2006). About 40% of costs in agriculture production are related to labour costs. Therefore it is the effective management of these costs that plays a vital role in the competitiveness of agricultural production which depend largely on an active labour force that is healthy, well trained, and adequately fed. But because of high level of illiteracy, hunger and poverty, these are endemic disease that affects the people and thereby reducing their productivity and capacity to execute certain activities. This does not make for sustainability. The human element in agriculture requires to be trained and managed well, so as to achieve the desired goal of sustainable food production overtime.

However, apart from the human element involved in agricultural production, there are institutional factors that need to be tackled if sustainability is to be achieved. These include research an extension training, land availability, credit institutions, cooperatives, marketing, and pricing policies (Ibeawuchi *et al.,* 2005). Sustainable resource management through agrarian science will lead to sustainable agricultural production as a component of economic development. This therefore requires good educational training and acquisition of skills in agriculture. Lack of adequate quantitative education, funding facilities and infrastructure scarcity of well trained experienced and highly competent workforce is in itself unsustainable. Agricultural resource management provide quality research, and extension services covering environmental planning, resource information, soil management, waste management and water management. Therefore manpower development and management is in itself a resource in agriculture that helps to manage most of the resources used in food production.

Biological Resources
The entire ecological system serves as a resource in agricultural production. The natural vegetation, forest, and animal resources constitute the main biological resources needed for effective agrarian revolution. However, because of poverty, hunger, and disease related factors, these resources are constantly being degraded. The destruction and loss of important vegetation and forest resources due to constant felling of trees for fuel and income have adversely affected agriculture and food production. Sustainable agrarian revolution cannot be achieved if important vegetation and forest resource such as wild life and game reserves are constantly being destroyed. To this end, alternative sources of income and fuel energy must be sought for and harnessed. Therefore, domestic energy which currently comes from fuel, wood must be obtained from more diverse sources.

Apart from these, important plant and animal species are constantly being lost annually through deforestation. Although deforestation leads to the loss of important plant species, wild life and game resources, it also promotes desert encroachment, wind erosion and global warming. Trees and vegetation act as wind brakes and prevent direct rays of the sun from reaching the earth's atmosphere.

Physical Resources

Physical resources such as water and soil are some of the limiting factors in intensive agricultural production. Land and water are two ingredients essential to sustaining agricultural operations. The duration and pattern of rainfall varies from one geo-ecological zone to another and in areas with lesser rain or totally dry in most part of the year, irrigation is used to manage the water for food production. In fact, irrigation systems are the lifetime for any agricultural operation through effective management. With efficient and trained labour force irrigation facilities well utilized can bring about sustainable agricultural production. However, the variability and unreliable rainfall, unpredictable periods of drought, very high precipitation, very high soil temperatures that damage crops, deforestation and desertification that adversely affect agricultural production and exacerbate the hunger and poverty situation is unsustainable (Ibeawuchi *et al.*, 2005).

The constant loss and destruction of top vegetable cover, degradation of rain fed croplands caused by wind and water erosion and the declining availability of water supplies can cause unsustainable resource management in agriculture. Apart from water resources, soil and soil resources are other limiting factors that affect food production. Low soil fertility and poor management to soil resources under high population pressure and intensive land use system make agriculture unsustainable and thereby increasing poverty gap and hunger. The proper management of soil and soil resources is very important if agricultural production must go on. However, soils have become so poorly managed that they are constantly being degraded through erosion, pollution and deforestation. Increasing intensity use and shortened periods of fallow have also played significant roles in reducing land productivity. Recovery of soil fertility can be achieved through proper organic manure practices and organic farming, agro forestry practices, covers cropping, inter cropping while erosion can be well managed or controlled through minimum tillage practice, intercropping with fast growing legumes and cover crops (Ibeawuchi and Ofoh, 2000).

Energy Resources

Energy is required to power some agricultural related facilities such as vacuum cooling plants, slaughterhouses and experimental stations. Energy sources such as thermal and hydroelectric and nuclear energy etc. are very important to agriculture. Constant energy is required to preserve farm inputs and outputs. A situation whereby there is constant interruption of power supply is unsustainable and can never reduce hunger and poverty. Energy is required to process agricultural produce and so add value to them. Hunger and poverty reduction could be possible if the poor have access to cheap, clean, and environmentally friendly constant supply of energy. Energy sources such as nuclear energy, hydroelectric, and fuel wood damage the environment. The scarcity of fuel wood at certain periods of the year create some additional problems for the women who are already over burdened with the generation of household income from self supporting activities such as crafts, trading and unskilled work at construction sites. This has the effect of reducing agricultural labour thus encouraging deagrarianisation and depeasantisation.

Women and children are especially pressed and suffer from various kinds of stress as more time is spent in fuel wood collection. This group constitutes the majority of the labour force needed in agriculture and as a result of this anything that can put additional burden to them will reduce food production and is highly unsustainable.

Agrarian Science and Practices for Sustainable Agriculture and Resource Management

The farmer at the centre of food production used the science and art of agriculture to maintain his farming environment by preventing erosion, maintain soil fertility, improve the genetic make-ups of plants and animals etc. for the feeding of the human population. These contributions will help in sustaining the environment through careful understanding and management of resources used in food production. However, climatic shocks such as food and drought and in many regions political instability have negative effect on the efforts of agriculture.

Agriculture or farming is a product of agrarian science. The question of sustainable agriculture using agrarian science has been an age long one bordering on the need to ensure a balance between food production and our environment. This is so, because environmental degradation and related process such as soil erosion depletion of minerals in the soil and loss of bio diversity etc. are known to have adverse effects on agriculture. While it has been agreed that environmental degradation such as deforestation, desertification water and atmosphere pollution are the results of poverty in developing nations of Africa,

Asia and Latin America, it is the result of wealth in the developed nations. Therefore sustainable agricultural development requires a comprehensive or integrated approach in the policies, planning, conservation, management, processing and utilization of natural resources. According to Rodale (1988), a sustainable system is one in which the resources used in production are managed in such a way that they are more or less self-generating and ensure continual improvement well beyond conventional expectation. However, the crucial role of political and administrative resources and policies, infrastructure, including markets, adequate input supplies and credit, institutions for education, research and extension, land tenure and appropriate laws and regulations in ensuring sustainability in agriculture must be put in place (CIGAR/TAC, 1988). Sustainable agricultural development cannot effectively be pursued sectional and in isolation. According to Okigbo (1991) it must be an integral component of an overall sustainable livelihood and development strategy, which gives priority to better management and conservation of resources so that their use in satisfying human needs minimizes damage to the environment. A sustainable agricultural production system therefore is defined as one, which maintains an acceptable and increasing level of productivity that satisfies prevailing needs and is continuously adapted to meet the future needs for increasing the carrying capacity of the resources base and other worthwhile human needs (Okigbo, 1991). Every aspect of sustainable agriculture deals with how to effectively use our resources so that the future will be assured. According to Okigbo (1991) sustainability can only be achieved when resources, inputs and technologies are within the capabilities of the farmer to own, hire, maintain and manage with increasing efficiency, to achieve desirable levels of productivity in perpetuity with minimal or no adverse effects on the resource based, human life and environmental quality.

To achievement sustainability thus requires rational and efficient management of natural resources including human. Majority of the earths mineral resources are regarded as non-renewable and so must be carefully managed. The extent, to which these resources can be effectively managed to reduce poverty and hunger, therefore depends by their amount, quality, location and availability in addition to the strategies and technologies adopted by the people to increase incomes. To this end, sectoral development activities in agriculture including fisheries and forestry, industrial and economic development must be combined in such a way as to minimize competition maximize compatibility and sustainability to reduce the adverse effect that they may have on each other (Ibeawuchi *at al.*, 2005).

Agrarian science overtime through proper research has evolved a lot of practices that is helping in increasing food production and effective resource management for sustainable agricultural production. Some of these practices include:

- Proper land use practices, which must include - zero and minimum tillage. In areas with fragile soils, mechanization or complete tillage that may lead to serious soil degradation must be avoided. Soil is the base for all agrarian practices care must be taken to manage it for sustainable crop production.
- Proper fertilizer use/application, to avoid nitrogen and phosphorus surplus into rivers and lakes.
- Low external input is encouraged.
- Pesticide uses are carefully applied so as not to pollute the environment to avoid destroying non target beneficial organisms.
- Adequate proper use of organic manure and materials is the basis for organic farming for sustainable food production. (Ibeawuchi *et al,.* 2006).
- Use of legumes and cover crops must be encouraged to enable the legumes enrich the soil with nitrogen fixed in the root nodules by bacteria especially in tropical soils (Brady, 1990).
- Research in agrarian science has evolved improved seeds and seedlings through plant breeding and genetic engineering.
- Research developed mechanical tomato harvester and agricultural scientists bred tomatoes that were harder and less nutritious (Friedland, and Barton, 1975). This had helped in improving feeding in rural areas.
- Agrarian science has developed art/science of manipulating crops to make changes in food composition. In use USA, a 50 years study on vegetable revealed that garden vegetables contain on average 38% less vitamin B2 and 15% less vitamin C (Davis and Riordan, 2004).
- Weed control-genetic engineering through the selection and breeding process developed an herbicide resistant gene that allows plants to tolerate exposure to glyphosphate which is used to control weeds in crops. Also a less frequently used but more controversial modification causes the plant to produce a toxin to reduce damage from insects (EC and CTA, 1999).
- Watershed management of rivers to prevent human influence on natural systems as practiced in Australia and America (Williams and Hunn *eds.,* 1982).

Environmental Problems

Looking at the impact of agriculture on the environment if the enabling resources are not well managed, agriculture may often cause environmental problems because it changes natural environment and produces harmful by-products. The environmental problem as a result of improper resour management in agriculture includes:

- Nitrogen and phosphorus surplus in rivers and lakes.
- Particulate matter including ammonia and ammonium off gassing from animal waste contaminating to air pollution and odour form agricultural wastes.
- Depletion of minerals in the soil through plant uptake and losses through bad farm practices.
- Consolidation of diverse biomass into a few species.
- Soil erosion caused by wind and water movements.
- Conversion of natural ecosystem of all types into arable and crop-able lands.

There are detrimental effects of weeds-feral plants and animals, pesticides, fungicides, herbicides, insecticides and other biocides. Soil desegregations caused by salination.

Government Policy

Government of countries of the world should encourage agriculture through enabling laws that are enforceable if not practicable for resource allocation and management in their respective countries. For instance in Africa as a result of structural adjustment performances, diversification out of agriculture has become the norm among rural population. This diversification takes numerous forms including migration especially younger people (men) and the sale of home making skills among women. Diversification offers many opportunities, but also brings high levels of financial and personal risks and threatens traditional agrarian and family values. If government policy is to support agrarian development, they must create jobs in the agricultural sector by the development of human resources, equipping young men and women with skills to work in agro-industries, farms, and other new agro-related environments. Government policy must include appropriate low cost ways of enhancing agricultural productivity deviating from the traditional ways or methods to attract young people.

There must be information from government about her policies through research and extension. Research plays a key role in generating, managing, and disseminating information for agricultural policy decisions. In Nigeria for example, the cassava initiative of the present Obasanjo led government has reached the rural people and it has given job to thousands of youths in the agriculture sector. These, therefore, must be backed up by the institutional capacity to perform the tasks needed to strengthen the agricultural sector for sustainable future.

Agricultural Research and Extension

Agrarian science derives its practices from research and extension. Through research and extension, innovations or research findings get to the rural people. However, the scientists in most developing countries usually focus on high value commodities rather than on crops, livestock and environments of greater relevance to the resource poor farmers. The research agenda is set by scientists, who may prefer solving scientific problems to addressing those faced by farmers and are rewarded by their output of journal articles rather than the impact of their technologies (EC and CTA, 1999). For the sustenance of agrarian families and values, at the planning stage of the research agenda, at a minimum there should be consultation with the resource poor farmers and those who serve them such as extension workers and NGO personnel. In the implementation stage, a whole range of participatory research tools should be available, from on farm trials managed by the researcher to experiments planned and managed by the farmers themselves.

The extension system must be more closely related to research, to ensure that the findings are disseminated widely. Rapid translation of research findings must be done in such a way that the extensionists and farmers can access, such as demonstration plots, simple manuals, and radio programmes (EC and CTA, 1999).

Government should budget handsomely for research and extension and it must be focused on the problem of hunger and poverty. Staple crops or those cash crops that can be grown by resource farmers, small livestock, poor soil, risky environments, and low levels of inputs should be made the focus and addressed through research.

Policy goals for effective agricultural policy implementation, there must be stability and continuity of government. In general, agrarian policy focuses on the goals and methods of agricultural production. At the policy level, common goals of agriculture include:

- Food safety - ensuring that food supply is free of contamination.
- Food security - ensuring that food supply meets the population needs.
- Food quality - ensuring that food supply is of a consistent and known quality.

The policy also looks at conservation, environmental impact and economic stability.

Conclusion

The hallmark of sustainable agriculture resource management is the building of sound manpower base that will utilize the resources and packages developed by agricultural research. Also, absence of disruption in government will help stabilize the polity and government policies with reference to agriculture can be effectively implemented to enhance and sustain agricultural production.

References

Anon, 2006. Agricultural Labour Management. University of California, 3800 Cornucopia Way. No. A/Modesto.

Brady, N., 1990. The Native and Properties of Soils. Macmillan Publishing Company New York.

Bryceson, D. F., C. Kay and J. Mooji (eds.), 2000. "Disappearing Peasantries? Labour in Africa, Asia and Latin America". London Intermediate Technology Publications.

Consultative Group on International Agricultural Research Technical Advisory Committee (OGIAR/TAC), 1988. Sustainable Agricultural Production, Implication for International Agricultural Research. Reme TAC Secretariat FAD.

Davis D. R. and H. D. Riordan, 2004. Changes in USDA Food Composition Data for 43 Garden Crops 1950 – 1999. *Journal of the America College of Nutrition 23:* 6, 669-682.

European Commission and Technical Centre for Agricultural and Rural Cooperation (EC and CTA), 1999. Reducing Poverty through Agricultural Sector Strategies in Eastern and Southern Africa. Summary Report of a Workshop, Wageningen the Netherlands, 23 - 25 November, 1998. pp 15-41.

Food and Agriculture Organization Statistics (FAOSEAT), 2002. Agricultural Information Centre - World Population Estimate. Rome 2002.

Food and Agriculture Organization Statistics (FAOSTAT), 2005. Agricultural Information Centre – Cross Domestic Product. Rome 2005.

Friendland, W. and H. Barton, 1975: Destalking in Wily Tomato, a Case Study of Social Consequences in California Agricultural Research. University of California at St. Cruz. Research Monograph 15.

Ibeawuchi I. I. and M. C. Ofoh, 2000: Production of Maize/Cassava/Food Legume Mixtures in South-eastern Nigerian. Journal of Agriculture and Rural Development I. 1: 1-9.

Ibeawuchi, I. I., Nwufo, M. I., Obasi, P. C. and Onyeka, Up 2005: Sustainable Agriculture as a Tool for Poverty Alleviation, a Review of Strategies for Crop Production in South-eastern Nigeria. Journal of Agriculture and Social Research (JASR). 5: 2, 11-19.

Ibeawuchi I. I., Onwereemal, E. U. and N. N. Oti, 2006. Effects of Poultry Manure on Green (Amaranthus cruentus) and Water Leaf (Talinum triangulare) on Degraded Ultisol of Owerri South-eastern Nigeria. Journal of Animal and Vet. Adv. 5: 1, 53-56.

New Partnership for African Development (NEPAO), 2002. Towards Claiming the 21st Century Annual Report, 2002. Midrad South Africa.

Overseas Development Institute (ODI), 2000. Natural Resource Perspectives. ODI, No. 52.

Okigbo, B. N., 1991. Development of Sustainable Agricultural Production Systems in Africa. Roles of International Agricultural Research Centres and National Agricultural Research Systems. 1- 61.

Rodale R., 1988. Agricultural Systems: The Importance of Sustainability. National Forum 68: 3, 2-6.

Williams, N. M. and E. S. Hunn, eds., 1982. Resource Managers: North America and Australia Hunter Gatherers: AAAS Selected Symposium, 67. Washington, D. C.

AGRARIAN SCIENCE FOR SUSTAINABLE RESOURCE MANAGEMENT

G. O. Iherijika

Abstract
The livelihood of the world's population depends mostly on natural resources: land forest, water, and the air we breathe. The continual degradation and diminution of these resources due to the cultivation of land threatens the economic and social security of individuals, communities and countries as well as the intricate web of ecological, social, economic and cultural relations that bind the global community together. To be productive, our resources must be managed properly, enriched and harvested responsibly. In most countries, agrarian science does not encourage investment in equipment, seeds and other inputs or the use of intense farming methods. Farmers suffer from lack of land due to land tenure system with associated pest and disease management problems like the ability to practice crop rotation and intense inoculums pressure at the margin of small field. When secondary implements are used, they cause problems of deforestation, which encourage desert encroachment. Soil compactness, increased erosion and leaching is thereby creating problems in natural resource management. This is coupled with exploitation of natural environment due to mining of solid minerals and crude oil and agro-infrastructural developments. Agrarian science should be practiced in such a way that natural resources as forest, plant, wild life and livestock can provide communities with sustainable livelihoods as well as environmental services like a forest's role in cleaning, recycling and renewing of the air and water that sustain human life.

Objectives of the research
The objective of this research is to investigate agrarian science and its associated problems, its effects on sustainable resource management and strategies for sustainable resource management.

Introduction
Much of the world's population depends on livelihoods derived from natural resources, land, forest, water and the air we breathe. The continued degradation and diminution of these resources due to cultivation of land threaten the economic and social security of individuals, communities and countries, as well as the intricate web of ecological, social, economics and cultural relations that

binds the global community (Peterson 1991). Crop produce yield by using the available nutrients in the soil and the cultivation of land in turn leads to the loss of soil nutrients. This is coupled with other field operations such as use of secondary implements which help to compact the soil, tillage, cutting down of trees, and general clearing as well as using chemicals for pest and disease control or as fertilizer or lime material. All these practices carried out in the process of cultivation of land, create problem in sustainable resource management and also affects yield in subsequent seasons.

As farmers lose the ability to produce crops and raise livestock due to environmental degradation and insecure land tenure, as communities depending on forests see them destroyed as well as previously productive areas by droughts, urban centres swell with millions of destitute migrants who join the growing ranks of the unemployment.

Agrarian science our natural assets

Natural assets are the wealth on which human well-being and -survival itself ultimately depend on proper handling of agrarian science. Cultivation of land should be carried out in such a way that in addition to production of high yield of crops and animals will transform our natural resources to sustain global economics viability and provide a healthy environment for future generation.

To be productive, our resources must be managed properly enriched and harvested responsibly. It should also be the kind of resource to which production processes can be add value. And it should be passed along in an ecological resilient and productive condition that ensures its viability for the next generation. The world is facing two series crises, the population crises and the environmental crises. Between these two, the environmental crises are more serious. Noibi (1990) and Neely et al. (1990) observed that the exploitation of the natural environment either in form of deforestation, mining of solid minerals and crude oil and agro-infrastructural developments has resulted in the creation of numerous problems which has adversely affected the natural resource management.

Many farmers in the less developed world cultivate small portion of land as a result of land tenure system and all their practices do not take recognizance of the effects of these practices to our resource management. More so, the local crops like peanut, maize, millet yield poorly and often fail due to erratic rainfall. Few farmers have the means to invest in making their livelihood more secure and productive. Until recently, migration used to be the only way out of this predicament. As people left, the life of the land ebbed away too. Farmlands languished and livestock dwindled for want of regular care and the village community lost its most active and dynamic members.

Problems of agrarian science

Agrarian science has problems such as difficulty in the control of pests and disease. In some cases the produce are not for sale, farmers are illiterate and crude tools are used but when secondary implements are used, they cause problems of deforestation, desert encroachment, soil compaction, increased erosion and leaching and all these create a very big problem in our natural resource management. In most areas, agrarian science's labour supply is erratic and unreliable and there is waste of cultivable land as there is no means of replenishing the soil and as such the farmers will migrate to cultivate a fresh land. Also the process bores down farmers due to the unchallenging of the works and there is no profit gained because the produce is not for sale. Agrarian science is mostly traditional and superstitious in nature. There are also low capital requirements, which can be provided by friends, money lenders and family members. There is also no record keeping done in most areas and as such no improvement crop cultivation because the time before maturity is too long.

Advances to agrarian science

An advance to agrarian science resulted to cash crop production, which is a more advanced stage of agricultural production. The farmers specialize in the cultivation of cash crop like cocoa, kola, rubber, citrus, groundnut, oil palm, etc. These crops are for export and they require processing into finished products before they can be consumed and the farmer sells the product in semi-processed form (Ogeva, 1975). Farmers are vital to the economics of developing countries, yet the yields they produce are low.

In less developed countries, agrarian science does not encourage investment in equipment, seeds and other inputs, or to use intensive farming methods. They suffer from a lack of land, with associated pest and disease management. Problems such as the inability to practice crop rotation and intense inoculums pressure are at the margins of small fields. Even farmers of cash crops have little influence on market prices. Agrarian science needs crop protection measures, which are cheap, simple, cost effective and sustainable. Appropriate strategies include the use of resistant cultivars, pest and disease exclusion through quarantine, cultural and chemical control.

Problems of agrarian science

Problems facing agrarian science in the world today are complex:

- Farmers have no money to invest in modern cost-effective technologies for crop production. Example, crop protection equipment like a spray pump may not be within their scope. Improved seeds must bring great advantages to convince a farmer to buy a pump.

- There farmers have very little land, in most cases less than 5 hectares, on which they can live on. They may not be in a position to adopt crop rotation if they are also to feed their families.
- The farmers have some unique problems because of their small fields. Examples, the inoculums from neighbouring crops are often abundant and there is generally much spread of inoculums between fields.
- Due to limited labour and funds, they cannot exploit intensive farming methods.
- The governments of the farmers have attempted to help farmers by introducing mechanization on a communal basis. This has not worked well basically due to management problems.
- Mechanization, if it is to work properly, should be used at almost all staged in the production process. This is not the case of agrarian science. Example, they may get a tractor to open up a large pieces of land and yet have to weed by hand and use human labour for crop protection activities.
- In the case of cash crops, the farmers play no role in pricing their commodities, both within and in particular outside their countries, which perpetuates their poverty.

The holistic approach to agrarian and resource management

The world is facing two serious crises: the population crises and the environmental crises. Between these two, the environmental crises are more serious (Noibi, 1990). Sunnet (1973) observed that the exploitation of the natural environment either inform of deforestation, mining of solid minerals and crude oil and agro-infrastructural developments has resulted in the creation of numerous problem which has adversely affected the natural ecosystem (plant and animal communities) of many parts of the world.

It is worthy to note that scientists have often been wrong concerning agrarian science and it is important that their limitation is recognised. We have more knowledge about microscopic matters, but local farmers probably know more where continued observation and knowledge of inter-relationships are involved. Green revolution agriculture has been effective mostly in flats irrigated lands where farmers can control the environment.

In contrast, most rainfall in tropical agriculture occurs in complex, diverse and risk-prone environments. Research stations even in such areas tend to be resource-rich and have led to only incremental gains for resource-poor farmers. Farmers should develop diverse cropping system in response to an uncertain environment.

As environmental and health hazards mounts as a result of heavy pesticide usage, liming and fertilizer application, these lead to deterioration in rural economics, a new holistic perspective emerges in food production. This involves dynamically evolving system in which widely divergent agricultural practices and conditions are evaluated, modified and verified so as to create a productive and continuing resource management. The concept is environmentally oriented, although people and technology are also given significant weight. For effective resources management, agrarian science should involve the use of components of related disciplines and concepts such as integrated pest management (IPM), farming systems approaches in production, research and extension, agro-forestry and agro-ecology (Dar, 1999; Gips, 1990).

There are great contradictions between forest predation and forest conservation, between world markets demand for indigenous products and the unavailability of financing for indigenous market activity between indigenous goals for self-reliance and the government legacy of indigenous isolation and paternalism. This is to add flexibility to laws that permitted only subsistence use of natural resource in indigenous territories. Thus, establishing indigenous rights to pursue sustainable resource management, as a viable economic activity and prescribe harvesting techniques designed to limit collateral damage, to biodiversity in the farm generally and in the vicinity of each felled tree.

Agrarian science should ensure that natural resources such as forests, plants, wildlife, land and livestock can provide communities with sustainable livelihoods as well as environmental services, such as a forest's role in the cleansing, recycling and renewable of the air and water that sustain human life.

When this approached is applied to communities that depend upon converting agrarian science into sustainable livelihood, it becomes a strategy for agrarian science for sustainable resource management in line with Boyce (2001); Boyce and Pastor (2001) as well as Boyce and Shelley (2003). Strategies for sustainable resource management in the hands of low-income individuals and communities can simultaneously advance the goals of poverty reduction, environmental protection and environmental justice. Thus contributing not only to increased income but also to non-income benefits such as health and environmental quality.

Conclusion

Land cultivations are faced with problems of lack of money to invest in modern cost effective technologies, land tenure, difficulty in control of pests and diseases, side effects of chemical such as pesticides, fungicides, fertilizers and liming materials on non-target organisms and plants as well as on the environment, deforestation, soil compaction, where secondary implement are used, increased erosion and leaching, thereby posing great problems in sustainable resource management. Approach to these problems should include use of components of related disciplines and concepts such as integrated pest management (I PM), farming systems approaches in production, research and extension, agro-forestry and agro-ecology.

References

Boyce, J. K., 2001. From Natural Resource to Natural Assets. *New Solution.* Vol. 11 (3), 266 - 278.

Boyce, J. K. and Shelley, B. G., 2003. Natural Assets: *Democratising Environmental Ownership,* Washington: Island press. 10 - 28.

Boyce, J. K. and M. Pastor, 2001. Building Natural Assets: New Strategies for Poverty Reduction and Environmental Protection, Political Economy Research Institute and the Centre for Popular Economics, University of Massachusetts-Amherst. 6 - 18.

Bunnet, R. B., 1973. General Geography Diagram. Longman London 40-96, Quezon City Philippines: Department of Agriculture, Bureau of Agriculture Research.

Dar, W. D., 1990. Basic Elements of the Farming System. Approach to Research and Extension, Dilman, 10 - 24

Gips, T., 1990. Breaking the Pesticides Habit. Alternatives to 12 Hazardous pesticides. Penang, Malaysia: International Organization of Consumers Union. 4 - 18.

McNeely, R. A. and T. B. Warner, 1990. Conserving the Worlds biological diversity. Washington D.C. and Gland, Switzerland: WRI: World Conservation International. 4 -15.

Noibi, A. S., 1990. Challenges of Environment Education in Nigeria School: Of Lawal, M. B. and A. A. Mohammed (eds). *Environmental education workshops and seminar proceeding.* Lagos. Redfield Nigeria Ud. 26 - 32.

Ogiera, E. B., 1995. History of Agricultural Development. Comprehensive Agricultural science. *A. Johnson publishers* Lagos, Nigeria. Pp. 16 - 1

Peterson, J. H., 1991. A Zimbabwean approach to sustainable Development a Community Empowerment through wildlife Utilization. *Harare: Centre for Applied social science,* University of Zimbabwe. 2 - 14.

THE CASE OF INDIGENOUS RESOURCE RIGHTS IN AKWA IBOM STATE, NIGERIA

Emmanuel O. Eyo

Introduction

The increasing need to enhance farm output culminated in Nigerian government pursuing policies that gives adequate consideration to the environment. Such policies brought about the introduction of the National Accelerated Food Production Program, the National Green Revolution Program, the Operation Feed the Nation, Agricultural Cooperative Development, the Agricultural Credit Guarantee Fund Scheme and the River Basin and Rural Development Program, among others. One common feature of these programs is that they encourage farming practices that ensures the biological communities maintain satisfactory level of interaction with the physical and chemical environment and thus enhancing the resource base of the ecosystem. Unfortunately, one thing that these efforts have had to grapple with is the culture, values and traditional practices, which vary according to ethnic decent in Nigeria. Invariably, in Nigeria and elsewhere, there are several traditional methods of farming that are still combined with expressions of spirituality, these external pressures not withstanding (Gonese and Tivafira, 2001; and Udoh, 1999). Two such ethnic practices that persist among the people of Akwa Ibom are the Market Day system of selling foodstuffs and the application of traditional practices on farms.

This paper assesses the effect of these two ethnic practices on the use of sustainable farm practices on farms and the output of the farmers in Akwa Ibom State, Nigeria.

On the concept of market days

In Akwa Ibom State the first day of the week, Sunday (called Obo in the local language), was reserved for the worship of the supreme being, the second day, Monday (called Edem obo) and the third day, Tuesday (called Fion aran), were reserved for the worship of god of the moon, the forth day, Wednesday (called Edere etaha), was a day for festivity, the fifth day, Thursday (called Etaha) and the sixth day, Friday (called Edemetaha), were set aside for environmental sanitation whereas the seventh day, Saturday (called Fionetok), was reserved for idol worshiping. Conditioned by the cultural activities of each community, the respective villages selected a day convenient enough to take their goods to the market. Such market days were named after the local naming of the day in each week. Hence at each day, some markets in some communities operated while others did not. However, in the entire state, different markets are used throughout the week such that person involved in itinerant trading had the opportunity to move their goods from market place to market place on the various days. Although today, there has been enough external pressure that has transformed some of these markets to daily markets, while weekly markets of the original traditional settings have remained alive in Akwa Ibom State.

On farm level traditional practices

There are many traditional believes that appear to impact negatively on agricultural progress. Many communities still sacrifice to the gods of the land before the beginning of every planting season, others engage in several other practices believing that agricultural output growth is dependent upon these traditional practices. In fact, every community exercise control over their cultural institutions, territories, languages, values, knowledge systems and practices. Such control commonly result from the supporting traditional believes of the people. Traditional believes and practices are essentially generally accepted ways of doing things that are guided by past believes and are usually associated with widely accepted rituals or other forms of symbolic behaviour. In fact, in farming communities where the control of indigenous resource rights is strong, farmers find it difficult to separate the spiritual practices from physical and mental phenomena, instead they consider them strongly as practices that must take place for their agricultural activities to thrive, (Balasubramanian et. al., 2003). Invariably, where agricultural practices are associated with traditional believes, these believe become an important parameter to consider if innovations are to be successfully introduced.

Materials and methods

This study was carried out in Akwa Ibom State, Nigeria. It lays between latitude 4° 33 and 7° 25" North and longitude 7° 25" and 8° 25" East; and occupies an area of 7,246.01 square kilometres. Akwa Ibom state has long rain periods (April to October) and temperature varies between 29° C and 34° C all through the year. Agriculture is practiced by about 80% of the population.

This study used principally primary data and involved two sets of respondents. The first set of respondents are crop farmers who operate in communities where the major weakly market day system are still being operated whereas the second set of respondents included crop farmers who operate in localities where farming activities are still tied up with same traditional practices. A total of 340 respondents were included in the study including 260 for market day effect and 80 for the traditional practiced effect. Data analysis utilized various statistical tools including the multiple regression analytical technique. In the empirical model, it is postulated that farm output is a function of the number of market days actually used by the farmer and the number of traditional practices in use. In the first empirical model the effects of the number of market days used by the farmers, on the output is tested whereas the second empirical model tests the effect of the number of traditional practices used by the farmers on output produced.

MODEL 1:

$Y = f(X_1, X_2, X_3)$
Where
Y = Output in grain equivalent
X_1 = Number of market days used by the farmers
X_2 = Labour in Man-days
X_3 = Farm size hectares.

MODEL 2:

$Q = f(X_1, X_2, X_3 + X_4 + X_s)$
Where
Q = Output in grain equivalent
X_1 = Annual income in Naira
X_2 = Farm size in hectares
X_3 = Labour in man days
X_4 = Management (proxy)
X_5 = Number of traditional practices used

In order to measure management, the proxy approach was used. This involves the selection and scoring of questions that portrays the nature of management capability of the farmer. In this study, a total of 30 questions were formulated and appropriate responses by the respondents were scored. An individual scored if the best option was selected; 1 if the least option was selected, and 2 if the middle option was selected and the total score was converted into percentage.

Results and discussion

This study included 340 respondents, 23.53% of the farmers who grow various crops in areas where traditional norms feature prominently in agricultural practices, while 76.47% produce various crop in localities where the market day system of distribution of farm produce predominate. The majority of the respondent were in their active years of age, male, married and had obtained at least secondary education. Table 1.0 shows that 37.06 of the respondents were in the 41-50 years age category, 29.12% were in the 31-40 years category,

18.24% were in the 21-30 years age category, and 15.38% were in the 51-60 years age category. Although the number of male respondents included in the study was more than the female, the female clearly outnumbered their male counterpart among the respondents studied to analysis the effect of traditional believe on outputs.

Table 1.0 also shows that 63.82% of the respondents were married, 52.35% were single, and 11.18% were divorced while 4.41% were widowed. Although a good portion of the respondents had between 4 - 7 children, none of the respondents among those studied for the effects of the market days on output had more than seven children. In fact, 10.65% of the respondents all among those studied for the effects of traditional believe on output had between 8 and 11 children. However, the majority of these respondents (58.53%) claimed that they had between 1 and 3 children and had attained formal education. Information available in table 1 shows that only 27.35% of the respondents had obtained no formal education where as 27.35% of them had attained primary education: 27.94% has attained secondary education and 7.36% had attained tertiary education.

Farming in this study area utilizes the traditional hoe, machetes and bush burning and the like. The soil fertility sustaining strategies adopted include mulching, the use of chemical fertilizer, farmyard manure and application of house-hold waste on farms. However, an assessment of the extent of use of each of these practices by the respondents reveal that only 41.8% of the respondents applied chemical fertilizer on farms whereas the others did not apply chemical fertilizer; only 20% claimed they applied farmyard manure to the crops and 23.2% claimed they applied house-hold waste to crops. (See: Appendix 1)

Table 2: Soil fertility sustaining practices

S/N	TYPE OF	FREQUENCY [n - 340]	PERCENTAGE
1	Use of chemical fertiliser	142	41.8
	None-use	198	58.2
2	Use of farmyard manure	68	20.0
	None-use of farmyard manure	272	80.0
3	Use of household waste	79	23.2
	Non-use of household waste	261	76.8

Source: Field Survey, 2005

Table 2 shows that the use of land management practices is below average despite the several programs of government. This is not surprising because the general notion is that traditional sacrifices and other practices are synonymous with high output.

Types of traditional practices in farms

This study revealed that crop farmers in that study area sacrifice goats to the gods before land clearing; engage the local masquerades not only to burn the trash after land clearing but also to dance around the farmland during planting. However, Table 3 shows that only 46.25% of the respondents claimed that they sacrificed goats, before land clearing, 28.75 percent of them claimed they use masquerades to burn the dried plant residue before planting, 11.25% agreed that they use the local masquerades to run around the entire farm land during planting, whereas only 5% of the farmers use the services of women to run around the farm in cases of pests attack, as this is believed to be the common remedy for outbreak of insect pest. (See: Appendix 2)

When one considers the extent of practices of these believes on farms, there appear to be great variations between groups. These variations in the proportion of respondents practicing each traditional believe on farm suggest that in the near future these believes appear to be gradually growing out of use. Already table 3.0 shows that at the moment 8.75% of the respondents are no longer using these believes on their farms.

Effect of the number of traditional practices on output

There is no gainsaying the fact that people with strong believe in the traditional value system give credence to the traditional practice in farms thereby ignoring the use of scientific packages, which are introduced in the existing agricultural programs to enhance the resilience of the ecosystem. To capture the overall effect of the number of traditional practices performed by the farmers on the performance of the farm, the multiple regression analysis was used. The regression model attempts to analyse the effect number of traditional practices observed by the farms (among other variable) on the quantity of farm output produced. The result of the regression analysis is shown below.

$$Y = -0.672 + 7.7E - 5\ X_1 + 0.91\ X_2 + 7.10E - 3\ X_3 + 0.73\ X_4 - 0.43 X_5$$
$$t_0 = -0.39;\ t_1 = 1.703^*;\ t_2 = 3.93^*;\ t_3 = 0.066;\ t_4 = 1.69;\ -t_5 = -3.005^*$$
$$R^2 = 41.2\%;\ \text{Standard estimated error} = 1.81;\ F\text{ - test} = 8.98^*$$
$$[^* = \text{Significant at 1 \%(two tailed test.)}]$$

Although the explanatory variables included in the regression equation explains only 41% of the total variation in the dependent variable, the result of the regression analyses clearly indicates that the number of traditional practices the farmers employ were negatively related to the quantity of output produced. Invariable the more the number of traditional practices embarked upon by the farmers, the lower the level of farm output produced. However, the annual income, farm size, labour supply and managerial ability of the farmers were positively related to the quantity of output produced. This implies that farming is such that farmers can only increase income by producing more farm output; farm output can only be increased by increasing the land area under cultivation with the use of more labour and application of good farm management practices.

The multiple regression analysis also shows that only the farm size and the number of traditional believes practiced on farms have had significant effects of the quantity of output produced by the farmers. Unfortunately, every unit increase in farm size resulted in a less than proportional increase in farmers' output and a unit increase in the number of traditional believes practiced on farms reduced output by 0.43 units. The other variables that were not significant in their effects on outputs produced by the farmers contributed less than proportionately to the increase in the quantity of farm products of the farmers. Suffice it to say, that traditional practices on farms influence the pace of application of sustainable farming practices thereby translating into low output for the farmers.

Effect on market day system of distributing of farm produce on the output farmer can produce

The multiple regression analysis also formed the basis for analysing the effect maintaining local market days on agricultural output of crop farmers.

$$Y = 0.19 \quad + 0.416 X_1 \quad + 0.621 X_2 \quad + 2.74 X_3$$

$t_0 = 0.095$; $t_1 = 0.671$; $t_2 = 2.084$; $t_3 = 3.23*$

$F = 8.234*$; Standard. Error of estimate $= 3.86$; $R^2 = 31\%$.

The result of the multiple regression analysis shows that the number of market days used by the farmers increase as output increases. More so, farm output increase with increase in labour and farm size. Although the R^2 shows that only 31% of the variation in output could be attributed to the independent variables used in the regression equation, only farm size had significant effect on the output and each of the variables contributed less than proportionately to the output produced. When one considers the magnitude of the coefficients, it is clear that farm size was the most important variable that affected farmers' output. On the other hand the number of market days used by the farmers to sell their farm produce increase with the quantity of farm output that is technically possible to produce. Invariably, for more farm output to be produced, the farmer must be prepared to transport the produce from one market place to another. In view of the difficulty the farmers are likely to face in such a situation, replacing the traditional weekly market system to the daily market system should be encouraged. This would be invaluable in reducing the risk of moving farm produce from market to market. More so, by dictating time of harvesting, market days stipulate the timing of cash inflow in farms. Using a one day in a weak for sales of agricultural produced may actually affect profit of the farmers. In particular, a farmer who is unable to move her perishable produce weakly from one market to market is forced to exhaust stock of perishable goods at reduced prices. The market day system also encourages wastage, since crops harvested in excess remains at the close of the market, unless sold. On the whole, the once-a-week selling of farm produce has overtime made farmers to plan production to suit the existing market day situation. This puts a check on the size of operation, how intensively land management practices are used and the quantity of output a farmer can supply to the market.

Conclusion

Every community has cultures having spiritual traditions and knowledge system and tries to protect their indigenous resource rights. Unfortunately this research suggests that the application of spiritual traditions and knowledge systems to agriculture seldom yield positive outcomes. Instead appropriate land management practices that are supposed to enhance the resilience of the ecosystem, and continued sustainability of the agricultural resource base are taken for granted would be invaluable in enhancing the resilience of the ecosystem and ensure sustainability in the agricultural resource base. The call by Norberg - Hodge (2005), for efforts to strengthen communities having these traditional practices is notable, but such communities must be encouraged to optimise the application of scientific practices introduced to enhance sustainability in the agricultural resource base.

Appendix, Table 1: General characteristics of respondents

S/N	Characteristics Of Respondents	Traditional Believes Practicing	Market Days Observing Group	Total
1	AGE			
	21-30	19 (5.59)	43(12.65)	62 (18.24)
	31- 40	21 (6.18)	78 (22.94)	99 (29.12)
	41- 50	27 (7.94)	99 (29.12)	126 (37.06)
	51- 60	13 (3.82)	40 (11.76)	53 (15.58)
	Total	80 (23.53)	260 (76.47)	340 (100)
2	SEX			
	Male	35 (10.29)	143 (42.06)	178 (52.35)
	Female	45 (13.24)	117 (35.41)	162 (47.65)
	Total	80 (23.53)	260 (76.47)	340 (100)
3	Marital Status			
	Single	9 (2.65)	61 (17.94)	70 (20.59)
	Married	35 (10.29)	182 (53.53)	217 (63.82)
	Divorced	21 (6.18)	17 (5.00)	38 (11.18)
	Widow	15 (4.41)	0 (0.00)	15 (4.41)
	Total	80 (23.53)	260 (76.47)	340 (100)
4	No. of Children			
	None	10 (2.94)	0 (0.00)	10 (2.94)
	1- 3	26 (7.65)	173 (50.88)	199 (58.53)
	4 - 7	35 (10.29)	87 (25.59	122 (35.88)
	8 -11	9 (10.65)	----	9 (2.65)
	Total	80 (23.53)	260 (76.47)	340 (100)
5	Education			
	No formal education	24 (7.06)	69 (20.29)	93 (27.35)
	Primary Education	32 (9.4)	61 (17.94)	93 (27.35)
	Secondary Edu ..	21 (6.18)	108 (31.76)	129 (37.94)
	Tertiary Education	3 (0.88)	22 (6.47)	25 (7.36)
	Total	80 (23.53)	260 (76.47)	340 (100)
6	Land Akquisition			
	Inheritance	77 (22.65)	121 (35.59)	198 (58.24)
	Purchase	3 (0.88)	139 (40.88)	142 (41.76)
	Total	80 (23.53)	260 (76.47)	340 (100)
7	Farm Size (ha)			
	1 and less	57 (16.76)	48 (14.12)	105 (30.88)
	1.1- 2.0	23 (6. 76)	69 (20.29)	92 (27.06)
	2.1- 3.0	0.00	121(35.59)	121 (35.59)
	3.1 and more	0.00	22 (6.47)	22 (6.47)
	Total	80 (23.53)	260 (76.47)	340 (100)

Source: Field Survey, 2005.

TABLE 3: Traditional practices in farm

TYPES OF RITES	FREQENCY	PERCENTAGE
Goat and sheep sacrificed before bush clearing	37.	46.25
Masquerades burn the dried residue before planting	23	28.75
Local masquerades run around the entire farm land during planning	9	11.25
Women dance around the farm outbreak of pest	4	5%
No traditional practices	7	8.75
Total	80	100.00

Source: Field Survey, 2005.

References

Balasubramanian, A. V., Vijaylaksmi, K., Sridbar S. and S. Arumugasamy. 2003. "Modern Dilemmas and Traditional Insights". Ancient Roots New Shoots. Vant Hooft, B. Haverkort, and W. Hiemstra (eds).

Udoh, J. A. ,1999. Agricultural Extension Development and Administration, KATZNS Publishers ltd., Nigeria.

Eyo, E. O., 2005. Agricultural Development in Nigeria: Plans, Policies and Programmes. Best Print Business Press, Uyo. Nigeria.

Gonese, C. and R. Tivafira, 2001. Eco-cultural Villages in Zimbabwe. COMPAS Magazine for Endogenous Development: Ancient Visions and New Challenges. No 4., March 2001. K. Vant Hooft, B. Haverkort and W. Hiemstra (eds).

Norberg-Hordge, H., 2005. Discussing Cross Cultural Images and Local Economies. COMPASS, No. 8., Feb. 2005, 8 - 13.

FIRE-WOOD FUEL PRACTICE IN BENUE STATE: IMPLICATIONS FOR LAND USE

Margaret O. Ode

Abstract
This paper is to identify fire-wood fuel practice and the problems associated with its use in Benue State with a view to determine an alternative that will enhance its effective use. Specifically the paper looks into the habitual way of using fire-wood fuel in Benue State, the environmental hazards associated with fire-wood fuel. The paper discusses the energy content of fire-wood fuel, fire-wood harvesting and the environmental impact including deforestation as a result of fire-wood fuel practice. Alternative and effective ways to enhance (improve) fire-wood fuel practice are suggested.

Introduction
Fire-wood is related to trees unsuitable for building or construction. Fire-wood is a renewable resource provided the consumption rate is controlled to a sustainable level. The use of wood as a fuel source for home heat is as old as civilization itself. Goodyer (1980) is of the view that prior to industrialization of societies, man-kinds' requirement for energy have been met primarily by muscular effort, fuel-wood, direct solar warming and punctuated occasionally by the use of coal. Wood burning is the largest current use of biomass derived energy (Wikipedea, the free encyclopaedia report). Wood can be used as a solid fuel for cooking or heating or occasionally for steam engines. Wood heat is still common throughout most of the countries of the world as well as in Nigeria and in Benue State in particular. Postel and Heise (1988) stated that more than two-third of all third world people rely on wood almost exclusively, even in oil-rich Nigeria. Akingbode (1993) observed that a winsome trend in the use of fuel-wood in Africa is that in the household sector. Akintoye (2002) found that nearly 80% of the energy flows were in the form of wood-fuel. The per-capital consumption of fuel-wood came to approximately 1.7m3/year. According to Reddy (1980) fuel-wood supplies roughly 80% of the useful energy in Southern India. Other sources of fuel include coal, oil or natural gas heating but too costly for even the average man to afford in Benue State.

Energy Content of Fire-wood
The energy content combustion of fire-wood depends mainly on how dry it is. "Green" wood is about 10mj/kg (mega joule per kilogram), air-seasoned wood is about 16kj/kg white km, while dried wood is about 19 to 20mj/kg (Wikipedia,

the free encyclopaedia report). The potential heat content per-kilogram of wood is roughly equal for all wood varieties. However, the heating potential of fire-wood per cubic meter or per log varies widely, depending on the species of tree from which the wood is cut and the density of the log. Generally, the harder the wood (which results from slower growth), the denser it is and the greater the account of biomass per unit volume. Such woods, when well-seasoned, produce hot, long-burning fires with practically no particulate emissions.

Fire-Wood Harvesting

Source fire-wood is harvested in purpose grown "wood lots", but in heavily wooded areas it is more usually harvested from natural forests. Deadfall that has not started to rot is preferred, since it is already seasoned. Standing dead timber (wood) is considered better still, as it is both seasoned and has less rot. Harvesting this form of timber (wood) reduces the speed and intensity of bush fires. Harvesting of fire-wood is normally carried out by hands, machetes and axes. Heavy fire-wood timber is harvested by hand with chainsaws. Therefore, longer pieces requiring less manual labour and less chainsaw fuel are less expensive. The prices also vary considerably with the distance from wood lots and quality of the wood.

The growth in popularity of fire-wood fuel in Benue State lead to the development and marketing of a greater variety of equipment for cutting and splitting wood. Today fire-wood fuel still continues to be used in both rural and urban areas where fire-wood is abundant. Many of the household in rural and urban in Benue State have no real alternative to fire-wood other than crop residues or cow dung's which are often unavailable. There is also scarcity of fire-wood and this causes social and economic difficulties in the area and people travelled long distance in search of fire-wood. The shortage of suitable fire-wood in some places has seen local populations damaging huge tracts of bush thus leading to further desertification. The combustion by product of wood burning is wood ashy which in moderate amount is also used as fertilizer and contributing minerals but is strongly alkaline. Wood ash is also used to manufacture soap.

Environmental impact on the use of fire-wood as fuel

Fire-wood consumption or heating has been singled out as a serious health hazard in many regions of the world (Wikipedes, the free encyclopaedia report). United Nations Institute for Social Development (UNISD) (1992) is of the view that biomass, used mainly by households in developed countries Like Nigeria, is the largest source of energy and efficiency in its use will be important in controlling air pollution. Depending on population density, topography, climate conditions and combustion equipment used, wood heating may cause serious air pollution problems especially particulates (Wikipedia, the free encyclopaedia report). Wood combustion is also known to release various quantities of toxic and carcinogenic substances. The conditions in which wood burnt will greatly influence the content of the emission, but in general, wood heating is not a healthy solution for residential heating. Fuel combustion (including fuel-wood) has been identified as one of the principal sources of atmospheric pollutions (Akintoye 2002). Fuel combustion releases nitrogen, sulphur in addition, about 594,000 tones of smoke particles were estimated to be emitted annually into the atmosphere from the burning of about 80 million cubic meters of fuel-wood (Freeman 1989, Akintoye 2002). This poses a number of health hazards.

Slow combustion stoves increases particulate production. The traditional stoves of three nude places and iron tripod perform inefficiently due to the absence of supply control and insufficient combustion. These stoves do not have any system of channelling heat flow and the escape of smoke. Therefore, households are exposed to a variety of health hazards such as bronchitis emphysema, lung cancer, blindness and so on (Akintoye 2002). It is essential therefore that in Nigeria and especially in Benue State, there should be improvement on the efficiency with which fire wood is produced and consumed in order to ensure sustainable developments.

Deforestation

Another environmental impact of fire-wood fuel is deforestation. It is the conversion of forested areas to non-forest. This meant conversion to grassland or to its artificial counterpart, grain-fields, urbanization and technological uses (Wikipedia Org/Wiki/Deforestation report). Food and Agricultural Organization of the United National (FAO) defined deforestation as the loss or continual degradation of forest habitat due to either natural or human related practices. That mining and petroleum exploration all contribute to cause deforestation. Natural deforestation can be linked to tsunamis, forest fires, volcanic eruptions, glaciations and desertification; the efforts of human related deforestation can be mitigated through environmentally sustainable practices that reduce permanent destruction of forests or even act as to preserve and rehabilitate disrupted forestland (Wang Honchang). The United Nations Research Institute for Social Development (UNRISD) (1992) also observed that deforestation include degradation that reduces forest quality such as the density and structure of the trees, the ecological services supplied, the biomass of plants and animals, the species diversity and the genetic diversity.

The term deforestation has also been used to refer to fuel-wood cutting, commercial logging, as the slash and burn technique, a component of some shifting cultivation agricultural systems. It is also used to describe forest clearing for annual crops, for grazing and establishment of industrial forest plantations. Deforestation causes has been practiced by human for thousands of years. Fire was the first tool that allowed humans to modify the landscape. Fire was probably used to drive game into more accessible areas. With the advent of agriculture, fire became the prime tool to clear land for crops and cattle farming (Natural History of Europe, 2005).

The short-sighted, marked-drive forestry practice is of the leading causes of forest degradation, the principal human-related causes of deforestation are agriculture and livestock grazing, urban sprawl and mining and petroleum extraction (William, 2003). Causes include demand for farm land and fuel-wood. Underlining causes include poverty and lack of reform. The largest cause of deforestation as of 2006 is slash-and-burn activity in tropical forests. Slash-and-burn is a method sometimes used by shifting cultivators to create short terms yields from marginal soil. When practiced repeatedly, or without intervening fallow periods, the nutrient poor soils may be exhausted or eroded to an unproductive state (Wunder, 2000). Slash-and-burn techniques are used by native populations of over 200 million people world-wide including Nigerians and people of Benue State.

Today growing world-wide demand for wood to be used for fire-wood or in construction, paper and furniture as well as clearing land for commercial and industrial development (including road construction) have combined with growing local populations and their demands for agricultural expansion and wood-fuel to end anger ever larger forest areas (Whitney 1996). One fifth of the world's tropical rainforest was destroyed between 1960 and 1990. Estimates of deforestation of the tropical forest for the 1990s range from ca 55,630 km^2 to ca 120,000 km^2 each year. At this rate, all tropical forests may be gone by the year 2090 (CFAN - CIDA Forestry Advisory Network, the Disaster of Deforestation and FAO - Forest Resources Assessment", 2005).

Today Nigerian's trees are vanishing at an alarming rate. And the World Bank figures have indicated that the country's losses in sustainable production of timber and fuel-wood from forestry resources, represents about US $750 million annually as a result of deforestation (Akingbode 1993). Nigerian Environmental Study Team (NEST, 1991), estimates the total annual consumptions of wood in Nigeria at about 50 to 55 million cubic meters of which about 90 percent is fire-wood. Annual deficit of fuel-wood in the Northern part of Nigeria is about 5 to 8 million cubic meters. The fuel-wood extraction rate in the Western part of Nigeria is about 3,885 times the rate of re-growth and almost ten times the rate of regeneration (NEST, 1991; Akintoye, 2002). The fire-wood consumption rate in the middle belt of Nigeria of which Benue State is inclusive is about 10 times the rate of re-growth and almost 20 times the rate of regeneration (NEST 1991). These figures give a rough idea of the magnitude of the deforestation and degree of severe population pressure on woody species in Benue State and Nigeria as a whole.

Environmental effort of deforestation

Deforestation is often cited as one of the major causes of the enhanced greenhouse effect. Trees and other plants remove carbon (in the form of carbon-dioxide) from the atmosphere during the process of photo-synthesis. Both the decay and burning of wood release much of this stored carbon back to the atmosphere, Akinloye (2002) asserts that overnight a stable forest releases exactly the same quantity of carbon-dioxide back into the atmosphere. Others states that nature forests are net sinks of carbon-dioxide (carbon-dioxide sink and carbon cycle), (The Danger of Deforestation Report).

Wild life

Some forests are rich in biological diversity. Deforestation can cause the destruction of the habitats that support this biological diversity, thus causing population shifts and extinctions.

Hydrologie Cycle and Water Resources

In general, trees and plants affect the hydrological cycle in significant ways including:
1. Their canopies intercept precipitation, some of which evaporates back to the atmosphere.
2. Their litter stems and trunks slow down surface runoff.
3. Their roots create macro pores - large conduits in the soil that increases infiltration of water.
4. They reduce soil moisture via transpiration.
5. Their litter and other organic residue change soil properties that affect the capacity of soil to store water.

    As a result, the presence or absence of trees can change the quantity of water on the surface in the soil or ground water, or in the atmosphere. This in turn changes erosion rates and the availability of water for either an ecosystem function or human services, (FAO/CIFOR report). The forest may have little impact on flooding in the case of large rainfall events, which overwhelms the storage capacity of forest soil if the soils are at or close to saturation.

Soil erosion

Deforestation usually increases rates of soil erosion by increasing the amount of runoff and reducing the profession of the soil from tree litter. This can be on advantage in excessively leached tropical rain forest soils. Forestry operations themselves also increase erosion through the development of roads and the use of mechanized equipment.

Landslides

The roots bind soil together and if the soil is sufficiently shallow, they act to keep the soil in place by also binding with underlying bedrock. The removal on steep slopes with shallow soil thus increases the risk of landslides, which can threaten people living near-by (categories: climate forcing agents report).

Controlling deforestation

New methods are being developed to farm more food crops on less farm land such as high-yield hybrid crops, greenhouse, autonomous building gardens and hydroponics. The reduced farm land is then depended on massive chemical inputs maintain necessary yields. In cyclic agriculture, cattle should be grazed on farm land that is resting and rejuvenating. Cyclic agriculture actually increases the fertility of the soil. Selective over farming can also increase the nutrients by releasing such nutrients from the previously invest subsoil.

Conclusion and suggestions for alternative and effective ways of managing fire-wood fuel

The Fire-Wood Fuel practice in Benue State was found to be ineffective due to the hazard impact it has on the environment. The smoke emitted by the fire-wood is hazardous to the environment and health of the people. Generally, the social economic aspect of fire-wood fuel practices in Benue State is due to poverty. Alternative and effective ways should be encouraged to enhance its usage.

The alternative and effective ways of managing fire-wood fuel will lesson the environmental hazards, associated with fire-wood consumption as well as ascertain saver ways of handling fire-wood fuel among the households in the urban and rural areas of Benue State.

- Fire places should be constructed on the ground and a smoke hole in the top of the tent or roof should be built to allow the smoke to escape by convention.

- In permanent structures, hearths should be constructed - surfaces of stone or another non-combustible material upon which a fire could be built and a smoke hold should be built in the roof through which smoke should escape.

- The development of the chimney and the fire place should allow for the effective exhaustion of the smoke. The building of masonry heaters or stoves should also help furthering capering much of the heat of the fire and exhaust in a large thermal mass, becoming much more efficient than a fire place alone.
- The metal stove should be used. The stove was a technological development concurrent with the industrial revolution. Stoves were constructed pieces of equipment that contained the fire on all sides and provided a means for controlling the draft - the amount of air allowed to reach the fire. Stoves have been made of a variety of materials. Metal stoves are often lined with refractory materials such as fire-brick, since the hottest part of a wood-burning fire will burn away steel over the cause of several years' use.
- Low pollution slow combustion stoves are a current area of research. An alternative approach is to use paralysis to produce several useful biochemical by-products and clean burning charcoal, or to burn fuel extremely quickly inside a large thermal mass such as a masonry heater. This has the effect of allowing the fuel to burn completely without producing emissions while maintaining the efficiency of the system.
- The technique of compressing wood pulp into pellets or artificial logs provides an excellent means of reducing emissions; not only is the combustion very clean (this of course depends on a well-designed combustion chamber and feeding system) but because of the increased wood density and reduced water content, the transport bulk is reduced by 30 to 70%. Thus the fossil energy consumed in transport is reduced and in fact represents a tiny fraction on the fossil fuel consumed in producing and distributing heating oil or gas.
- Hard woods, when well-seasoned produce hot, long-burning fires with practically no particulate emissions - so hardest, most dense wood should be most desirable for fire wood.
- Techniques to increase wood density (such as palletising or compressing of wood pulp into logs) can increase the heat content (per cubic meter) dramatically. At the same time, if the moisture content of this type of fabricated wood fuel is maintained at a consistently low level (typically 8 - 10%), the heating value is maximized and very high combustion efficiencies are possible. This in turn provides very clean burning-low emissions, little ash production and a minimum of soot and deposits in the combustion chamber and chimney.
- Efforts should be made to stop or slow down deforestation, because it has long been known that deforestation can cause environmental damage sufficient in some cases to cause societies to collapse, e.g. in Tonga, where the Paramount Rulers Developed Policies designed to prevent conflicts between short-term gains from converting forest to farm land and long-term problems forest loss would cause, whilst during the seventeenth and eighteenth centuries Tokugawa

Japan, the shoguns developed a highly sophisticated system of long-term planning to stop and even reverse deforestation of the preceding centuries through substituting timber by other products and more efficient use of land that had been farmed for many centuries. Management of forest lands can be sustainable to maintain both forest cover and provide a biodegrable renewable resource forests are also important stones of organic carbon, and forests can extract carbon-dioxide and pollutants from the air, thus contributing to biosphere stability and probably relevant to the greenhouse effort. Forests are also valued for their aesthetic beauty and as a cultural resource and tourist attraction.
- New methods are being developed to farm more food crops on less farm land, such as high-yield hydrides and hydroponics. The reduced farm land is then dependent on massive chemical inputs to maintain necessary yields. In cyclic agriculture, cattle are grazed on farm land that is resting and rejuvenating. Cyclic agriculture actually increases the fertility of the soil. Selective over farming can also increase the nutrients by releasing such nutrients from the previously inert subsoil. The constant release of nutrients from the constant exposure of subsoil by slow and gentle erosion is a process that has been ongoing for billion of years. Slash-and-burn agriculture has recently needed re-evaluation as it appears to be more sustainable than originally believed.
- The government of Benue State and Nigeria as a whole should make a policy whereby every able-bodied citizen should plant at least five trees per year in the forest services. If this method is adopted, billions of trees should be planted in Nigeria every year. Increasing consumer demand for wood products that have been produced and harvested in a sustainable manner will be causing land owners and forest industries to become increasingly accountable for their forest management and timber harvesting practices.

References
A Natural History of Europe - 2005 co-production including BBC and ZDF.
Akingbade, Tr., 1993. Africa in Energy Crisis, Lagos. African Sc. Monitor. Vol. 3 No. 13
Akingtoye, M., 2005. Fuel consumption patterns of Households in Ondo State and strategies for its effective management.
Categories: Climate Forcing Agents/Environmental/Forestry,
CFAN - CIDA Forestry Advisory Network, 2006.
FAO., 1973. A wood consumption survey and timber trend in Gambia, Rome, FAO.
FAO/CIFOR Report, Forest and Floods Drowning in Fiction or Thriving Facts?
FAO - Forest Resources Assessment, Change in extent of forest and other wooded land 1990 –2005.

Freeman, B., 1989. Environmental Ecology. California, Academic Press Inc.

Goodger, E., 1980. Alternative fuel, chemical energy resource trees, Landon, Macmillan.

Natural, History of Europe, 2005. Co-production including BBC and ZDF.

Nigerian environmental study/Action Team, 1991. Nigerian's Threatened Environment, A National Profile, Ibadan, by Intec Printers limted.

Postei, L. and S. Heise, 1988. Energy and Agriculture in the Third World, Cambridge, Mass, Ballinger Publishing Co.

Reddy, A. K., 1980. Rural energy consumption pattern: A Field Study, Bangalore, Cell for Application of Science and Technology.

Retrieved from http://en. Wikipedia. Org/wiki/Deforestation.

UNRISD, 1992. World Report, Washington D.C., United Nations.

Wang Hongchang. "Deforestation and Desiccation in China: A Preliminary Study".

Whitney, G. G., 1996. From coastal wilderness to fruited plain; A history of environmental change in temperate North America from 1500 to the present. Cambridge University Press. ISBN 052157658x.

Wikipedia, the free encyclopedia .Jump to: Navigation search fire-wood stacked to dry.

William, M., 2003. Deforesting the Earth, University of Chicago Press, Chicago. ISBN 0226899268.

Wunder, S., 2000. The Economics of Deforestation: the Example of Ecuador. Macmilla Press, London. ISBN 033371468.

MANAGING AFRICA'S NATURAL RESOURCES TO MITIGATE LAND DEGRADATION
A SOIL SCIENCE PERSPECTIVE

Joshua O. Ogunwole

Abstract
From the seventies, sub-Saharan Africa (hereafter referred to as Africa) has been the only subcontinent of the developing world where per capita food production has been declining. The need to improve the living standards in the subcontinent is putting enormous pressure on the natural resources. Soils are eroding, forests are shrinking and aquifers are being depleted. These have posed a great challenge to soil science and natural resource managers in the subcontinent. Land management practices were base on agricultural research built on models of the previous colonial rulers, which are no longer sustainable. In the past, farmers have succeeded in maintaining a certain agro-ecological equilibrium through the indigenous cropping pattern and land use practiced. Current effort to mitigate land degradation is to use our understanding of science to build on this indigenous practice and in an integrated manner. A principal strategy to restore degraded soils and ecosystems is through development and promotion of biologically stable and efficient soil management practices that restore and maintain soil fertility along with sensible management of water resources.

Introduction

Agriculture dominates the economy of Africa, accounting for more than 70% of the workforce and over 25% of the Gross Domestic Product (GDP). Many African countries, despite their generous endowment with natural resources, are going through very severe food and economic crisis. The demographics of Africa are changing, causing more land to be converted into cropland and bringing natural ecosystems into cultivation. In spite of efforts to reduce poverty, the continuing high fertility rates in Africa with higher food and water demands is causing more human influences on natural resources. High food production is being achieved with high inputs of water, fertilizers and energy to the soil causing increased pressure on the soil resources.

The intensification of land use systems in Africa has adversely affected her natural resources of which land degradation, forest depletion, and looming water scarcity are among the most important. Past mismanagement of these natural resources in agricultural production has reduced the sustainability of most of

Africa's ecosystems. The prospect of negative effects on the natural resources from agricultural technology is fast becoming a major issue of concern for natural resource scientists. Sustainable management of natural resources is an important pre-requisite for improved agricultural productivity and an ecological balance. As agriculture seeks to be more sustainable in Africa, it will need to be ever more science-based than before. This is where soil science comes in.

As natural resources in much of Africa have degraded over the years, the principal challenge to soil science and soil scientists is to manage and sustain the natural resource base by containing and possibly reversing the degradation processes of these resources. This is with the aim of rejuvenating the natural resources for meeting the ever increasing food demands of Africa's high population rate while protecting the environment (Rashid, 2006).

Land and environmental degradation

The term 'land' is described as a complex of soil, climate and biotic resources within major terrestrial ecosystems. Land quality refers to the capacity of all the components (i.e., soil, climate and biota) to produce economic goods and services depending on the land use and management. The loss therefore, of these intrinsic qualities of land due to natural or human induced processes that decrease the current or future capacity to support life-sustaining functions is referred to as land degradation (INRA, 1995). Land degradation leads to adverse changes in intrinsic qualities of its components such as climate, soil, water, vegetation and fauna resulting in declining productivity and reduction in environmental regulatory capacity. In Africa, soil erosion is fast exceeding soil formation, carbon emissions seem to exceed carbon fixation, losses of soil nutrients exceeding additions of nutrient to soil and forest destruction exceeds forest regeneration. One of the causes of this accelerated degradation of land resources is the pressures for conversion to agricultural use and human settlements of land with agricultural potential. The increasing claims of agricultural land for non-agricultural uses are mild when compared with those placed on water resources. While the increasing population requires more food, availability of water for irrigation is decreasing. Most human-induced land degradation is caused principally by inappropriate agricultural activities, overgrazing, over-exploitation and deforestation.

Development of sustainable natural resource management to limit land degradation and its consequences is indeed a great challenge to natural resource scientists. Sustainable development means not only caring for ourselves today but also leaving the subcontinent a better place for posterity. Exploring the potential for doing so requires the systematic application of human know-how for natural resource and ecosystem management to harness the desired benefits. Agricultural technology will need to adjust rapidly to address increased scarcity of water, soil nutrient depletion and forests recession. Efforts must be put in place to develop technology with active interaction among and between scientists and farmers to maximize utilization.

Water scarcity and environmental challenges

Water is a vital resource in any agricultural development. Hence, agricultural water use is the main one among all water users. This is because it has the largest water demand. Despite the fact that this use plays an essential role in Africa's food and fibre supplies, it provides for mitigating poverty and produces a major source of income, it is unable to compete economically for scarce water among other competitors (Pereira, 2005).

Rain is the primary source of agricultural water in Africa. It is subjected to extreme fluctuations. Inter- and intra-annual variability in rainfall is perhaps the key climatic element that determines the success of agriculture particularly in the drier parts of Africa. This erratic behaviour of rainfall in the subcontinent results in a high coefficient of variability (CV). Rainfall varies on all time scale: annual, monthly and weekly. The CV increases substantially from the annual to the weekly period (Sivakumar, 1997). Variability in rain fall also exists in space. Lawson and Sivakumar (1989) showed that CV for July and August rains is much less at the semi-arid Ouagadougou (Burkina Faso) than in the humid Ibadan (Nigeria). The subcontinent can therefore be said to be characterized by a steep gradient in annual rainfall from as low as 100 mm in Niamey (Agnew, 1982) to more than a few metres in Zaire. Rainfall variability also dictates to a large extent the distribution and extent of climatic zones in Africa, these include desert (29%), arid (17%), semi-arid (8%), dry sub-humid (11%), moist sub-humid (20%) and humid (14%). From this distribution it is obvious that rainfall is, in most parts of Africa, for at least part of the year, insufficient to grow crops and rain fed food production will be affected heavily by the annual rain fall variation. Combating water shortage requires that particular attention be given to efficient management of irrigation, water minimizing, water wastes and managing water quality. In addition, there is a need to improve the region's capacity for climate (particularly, rainfall) monitoring and prediction.

Soil fertility depletion

The low agricultural productivity in Africa is strongly related to the low quality of the soil resource base. Most of the soils of Africa are characterized by inherent or induced deficiencies of the macro- and/or micro-nutrients. Over the years, the resource-poor farmers of this region have removed large quantities of nutrients from their soils without returning them as manure or mineral fertilizer in sufficient quantities. The introduction of improved crop varieties that increase nutrient removal without compensating soil amendments has further heightened the threat of these deficiencies. Hence, soil fertility depletion has been described as the fundamental cause of low per capita food production in smallholder farms in Africa. Declining fertilizer use caused by elimination of subsidies as well as unsuitable crop rotation has resulted in negative nutrient balance in most of the soil of the subcontinent. Within the last 3 decades, average rate of soil fertility depletion was calculated as 22 kg nitrogen, 2.5 kg phosphorus and 15 kg potassium per hectare of cultivated land (Sanchez, as quoted by Spore, 2003).

Accelerated soil erosion is another major cause of soil fertility depletion on agricultural lands in many parts of Africa. Severe soil erosion has been reported in the highland areas of Ethiopia and Kenya. Soil loss in Rwanda has been estimated at an average of 5 kg/ha/yr. In south-eastern Nigeria, gully erosion is a problem, advancing between 20 to 50 metres yearly (Lal, 1995). Countries like Lesotho, Zimbabwe, Tanzania, Botswana and Burundi have been designated erosion-prone areas.

Improvement and maintenance of soil quality in agricultural land of Africa is vital if agricultural productivity and environmental quality are to be sustained for posterity. The way to restore and replenish fertility of these soils lies along the path of management practises that mimics the natural ecosystem from which current agricultural systems develop.

Dwindling forest resource
Forests are an extraordinary resource that cater for both the needs of man (as habitats for biodiversity) and that of nature (as major carbon sink). Africa has a forest cover of about 520 million hectares, out of which 48.3% is located in the humid Central Africa (FAO, 1999). Demographic changes are currently placing more pressure on the forest resulting in a continuous decline in forest cover. An estimated 2 million hectares of forest land is lost annually in West and Central Africa.

One major cause of dwindling forest resource is the expansion and intensification of agriculture which has also contributed to intensified pressures on the forest resource. In many forested ecosystems in Africa, forests continue to fall with clearly driven in part by demand for agricultural land. Changes in forests of Africa are usually dominated by transition from closed forests, through intermediate stages of depletion, to shrub and short fallow. This is an indication that the extension of small holder slash-and-burn agriculture under increasing demographic pressure is a primary cause of the dwindling forests. It is only by fostering technologies integrating both trees and crops on the same piece of land that the problem of dwindling forest resource could be alleviated. For resource-poor farmers of Africa, only technologies that add value to forests will be easily adopted.

Traditional farming systems

As earlier mentioned, there is a large climatic variation in the subcontinent, both from east to west and from north to south. Its characteristic climate is one of warm rainy season; cold, harmattan and hot dry season. Although a rich variety of sub-climates prevail in the African sub-continent, it can be broadly divided into three main agro ecological zones based on rainfall amount and length of growing season (LGS).

a. Semi-arid: This zone stretch from the middle veldt parts of South Africa across to southeast of Zimbabwe. Extensive parts of Botswana which continues to Namibia belong to this ecological zone (Nandwa, 2003). In West Africa, this zone forms a continuous band from Senegal and Gambia in the West to Chad in the East. It is characterized by rainfall limits corresponding to 250-900 mm and 75-119 days LGS. In this zone, rainfall exceeds potential evapotranspiration from two to five months annually. This zone is also referred to as Sudano-Sahelian zone *(FAD,* 1999) and has received lot of attention in the last 3 decades, because recurrent drought and successive crop failure have led to declining national income and decreasing per capita food production (Sivakumar and Wallace, 1991). Any agricultural development effort in this zone, that fails to take into consideration the poor and harsh environment in which crops are grown, will only have a short-term benefit.

b. Moist savannah: Coastal lowlands of Kenya, Madagascar, Uganda, Ethiopia, Eritrea, Rwanda and Burundi. The coastal savannah of West Africa coast stretching from Togo to Ghana and the Guinea savannah of Nigeria, Cote d'Ivoire and Guinea falls into this zone. The sub-humid Southern Africa is covering Tanzania, Zambia, Madagascar, Namibia, Angola and Lesotho. It is characterized by annual rainfall range of 900 - 1500 mm with LGS of 120 - 270 days. Vegetation here includes the Northern and Southern Guinea savannah and the Derived savannah. Blessed with abundant solar radiation, this zone has a high agricultural production potential.

c. Humid zone: This zone is characterized by annual rainfall in excess of 1500 mm, with more than 270 days LGS and rainfall in excess of evapotranspiration in virtually all the months of the year. This zone includes all the coastal swamps, mangroves, fresh water swamps and tropical rainforests of the subcontinent. Okigbo (1996) reported a great modification of the rainforest in some parts of the subcontinent, from the climax to simplified vegetation types such as forest savannah mosaic (Derived savannah) and secondary bush by human activities like farming, grazing, biomass burning, construction, etc.

With the exception of very few areas in Africa where farming methods are patterned after the modern systems of developed countries, multiple cropping constitutes a major component of the traditional farming systems. In the Africa subcontinent, standard or rigid patterns of intercropping are not clearly defined. Different farmers grow crops in a number of different combinations; however dominant intercropping pattern exists within any one agro ecology. Farmers generally tend to adapt by making sure that the less nutrient demanding species are frequently relay intercropped into the high nutrient demanding species.

Four major systems based on the dominant crops can be distinguished in the Africa sub-region. In the humid zone and some parts of the Derived savannah, the traditional cropping systems are based on yam, which is normally planted in the first year after bush clearing. Yams are often intercropped with cassava, maize, vegetables, bananas and pineapples. Tree crops like oil palm, rubber and cacao are also cultivated. In this zone, tropical hard woods are a valuable resource where the forests remain (Okigbo, 1996). The maize-based cropping system predominates in the moist savannah. It is usually intercropped with cassava, yams and cowpea. Banana, pineapple, sugarcane and rice are also cultivated here. However, the maize/cassava intercrop has been identified as a very productive combination and grown extensively in this zone. The heavy tse-tse-fly infestation in the moist savannah limits the exploitation of live stock potentials. The cropping systems typical of the semi-arid zone are the sorghum and millet-based systems. Major crops in these systems include groundnut and cowpea. Cotton, tobacco and sweet potato are also some of the cultivated crops in the zone. Along with arable crop production in this zone, livestock (cattle, sheep and goats) contribute to the farmers' welfare and income as they improve tillage through traction and fertility through manure, exploit crop stoves and by-products, produce live weight, milk, hides and skin for sale and 'live bank' when cash is needed. Hence, this zone has a high potential for livestock production. In the wet season, there is usually a major influx of nomads to this zone.

Shifting cultivation which involve 2 - 3 years of cropping and 10-years fallow without any farm power, along with bush fallowing or land rotation are widespread traditional methods of soil fertility maintenance in Africa. These management practises are biologically stable and efficient soil fertility restoration practises, since they mimic the natural ecosystem from which the cropping systems developed. Demographic and economic changes have, however, caused a widespread disappearance of these soil fertility restoration practices and cultivation has expanded into marginal soils. The result of all these are the systematic degradations of many land holdings and declining yield of crop plants.

Science-based natural resource management strategies

A prosperous agriculture sector in Africa is important for future development without which poverty cannot be reduced, natural resources cannot be managed in sustainable manner and, food security cannot be assured. Science based improved technologies can be the driving force for sustainable agricultural systems that provide more food while protecting the environment. It is now well recognized that the past dependence on agricultural research which has been built on the models of the previous rulers, contributed to raise production and productivity, but it had some undesirable effects, e.g. the dominance of inorganic fertilizer, discouragement of mixed cropping, emphasis on engineering rather than biological approaches to soil stabilization, etc.

Strategies to effectively manage natural resources to mitigate land degradation will involve three overriding objectives. The first is to seek or develop systems and system components that promote and maximize soil cover. Vegetative cover is vital to minimizing erosion and conserving soil moisture. The second objective is the restoration and maintenance of soil fertility while, the third will be the sensible management of water. These strategies must be guided by an integration-oriented paradigm for natural resource management, if sustainable agricultural development is the goal. These management strategies must be easily affordable and not be too labour intensive as women and children are the major source of labour in most agricultural systems of Africa.

Cover cropping

Appropriate management of ground covers can be effective in soil conservation, weed control and their organic mulch-dead or -alive can bring about long-term soil improvement and suppress weeds. Cover cropping has become an effective alternative to the bush fallow system. It can increase SOM beyond what is possible by simply leaving crop residue in the field. Herbaceous legumes are also useful in managing bush fallow, especially to suppress noxious weeds such as *Imperata cylindrica*. In parts of Africa, where shortage of arable land is acute, *Mucuna pruriens* var. utilis can be sown to reclaim land that was previously abandoned to this weed species.

Screening results have shown that *Desmodium heterocarpon* and M. *pruriens* can cover 50% of the ground within 8 weeks after planting and by 24 weeks, M.

pruriens, Centrosema pubescens and *pueraria phaseoloides* can cover 100%. The use of cover crop can serve a multitude of functions, such as erosion protection, nitrogen fixation, addition of crop residue to build SOM contents, and weed control. Cover cropping with mainly leguminous species has been reported to be more efficient and require less time in restoring soil fertility than natural fallows (Lal and Okigbo, 1990). In the moist savannah (Odunze *et al.,* 2004) showed that sole legume planting for two years restored fertility status of soils and enhance the soil organic carbon and total soil nitrogen contents, which resulted in increased maize grain yield than under the sole maize despite addition of 120 kg nitrogen (N) fertilizer to the sole maize.

Agro-forestry systems

The challenges to sustainable agricultural development in Africa are multifaceted, but it is clear that agro-forestry is one important option that can successfully address the challenges of food security, poverty reduction and environmental protection. Agro-forestry, which is the growing of trees and crop together, was developed on the observation that in the process of "slash-and-burn", African farmers usually deliberately retain certain tree and shrub species in their crop production systems, because they recognize that these tree-or-shrub species have long-term beneficial effects. Agro-forestry systems seek to fulfil the promise of trees and shrub species for helping to reconcile the goal of more intensive food production with that of maintaining long-term viability of farming systems. As a system that also allows multi-storey cultivation, rejuvenation of natural trees along side annual crops and economic trees, agro-forestry system can be a good alternative to the shifting cultivation (slash-and-burn) system.

Alley cropping is one agro-forestry system that imitates and improves upon the bush fallow system. It indicates a potential for developing a stable and productive system with a tree or shrub component which recycles plant nutrients and provides materials for mulch and support to twining crops like yam in the humid zones. *Leucaena leucocephala* planted in rows 4 m apart and intercropped with maize produced substantial amounts of pruned top dry matter and nitrogen yield, which benefited the associated crop and improved soil fertility (Kang *et al.,* 1995). The use of *Leucaena* tops maintained maize grain yield at a reasonable level with no N inputs on a low-fertility soils of the humid forest ecosystem. Likewise, tops of *Azadirachta indica* and *Parkia biglobosa* have been reported to improve soil fertility and yield of maize in the moist savannah (Uyovbisere and Elemo, 2002).

Tillage and residue management

The way to dramatically reduce the effect of accelerated soil erosion in Africa is through the zero tillage system. The tillage method ensures the soil is covered with crop residue, reduce soil disturbance to almost zero and attempts to maximize the number of days in the year when living roots grow in the soil. Over time, soil quality will improve in a zero tillage system through increased SOM, improved soil structure and water infiltration. Sustainable agricultural techniques like zero tillage enables farmers to preserve their land while ensuring regular and plenty harvest. This is because zero tillage can cut cost significantly in areas of labour and fertilizer requirement. Since land preparation is no longer necessary, cost of machinery and the fuel needed to power it is also saved. Zero tillage encourages high biological activities thus minimizing compaction and the organic matter that builds up in the soil traps carbon, preventing it from escaping into the atmosphere in the form of carbon dioxide, thereby helping to mitigate green house effect.

Biological nitrogen fixation approach to soil fertility improvements

Biological nitrogen fixation is an important and integral component of sustainable agriculture in the African farming systems. The capacity of legumes for symbiotic fixation of de-nitrogen is the major factor responsible for improving nitrogen fertility in soils. Legume inoculation with rhizobia is a mature agricultural bio-technology, however, rudimentary inoculation practices of moving soil from field previously cultivated with well-modulated legumes are also effective. Grain legumes such as soybeans and cowpeas, along with herbaceous legumes like *Centrosema pascuorum* and *Lablab purpureus* are planted in rotation with maize in West Africa. Adeboye *et al.* (2006) reported soybean and *Centrosema* to increase soil organic carbon and total nitrogen content in an alfisol of the moist savannah zone of West Africa.

Leguminous tree fallows of several species of *Sesbania, Tephrosia, Crotolaria* and *Cajanus* have been reported as possessing potential to accumulate 100-200 kg N ha-[1] in 6 - 24 months in their above- and below-ground biomass (Sanchez, 2001). The remaining biomass nitrogen from these trees can be incorporated into soil after the trees are removed for firewood. In the moist savannah zone of Southern Africa, two years tree fallow of *Sesbania sesban* and *Tephrosia vogelii* was reported to replenish soil nitrogen to levels sufficient enough to grow three subsequent high-yielding maize crops in nitrogen-depleted but phosphorus-sufficient soils (Mafongoya *et al., 2003).*

Manure management for cropping

One conventional method of improving productivity of degraded soils in Africa is through organic manuring in the form of livestock manure, crop residues and green manures. The beneficial role of these organic amendments cannot be over-emphasized. Organic manure (particularly, livestock manure) have the capacity to provide nutrients, especially N, P and potassium (K), increase cation exchange capacity and improve soil hydraulic conductivity. Before the advent of inorganic fertilizers, these organic manures constitute the principal source of nutrients to crops and are used in maintaining soil fertility.

In the West African semi-arid, the application of livestock manures takes several forms. Farmers may gather manure from stalls, transport it and hand spread it on the fields. This is recommended for poultry litter. However, for cattle droppings, corralling their animals on their fields overnight during the dry season returns both manure and urine to soils, resulting in greater crop yield and improved SOM and total soil N. Corralling cattle for 1, 2 and 3 nights have been reported to give manure application of 3, 7 and 10 tons ha-^{1} respectively (ICRISAT, 1991). Adeoye (1986) showed that application of cattle manure from 0 to 2.5 ton ha-^{1} increased organic carbon by 100%, reduced soil bulk density while, increasing mean weight diameter and available water (Table 1).

Table 1: Effects of cattle manure on properties of a loamy sand at Samaru, northern Nigeria.

Soil properties	Levels of cattle manure application (kg h^{a-1})		
	0	2500	5000
Organics carbon (g kg-^{1})	1.80	3.80	5.10
Soil surface bulk density (Mg m^{-3})	1.59	1.53	1.49
Mean weight diameter (mm)	0.41	0.46	0.56
Available water (mm)	27.30	30.30	37.80

Source: Adeoye, 1986.

Inorganic fertilizer application is one simple way to addressing the problem of declining soil fertility. However, many African farmers have difficulty raising enough cash to purchase fertilizer. Throughout most of Africa (particularly, East and Southern Africa), there has been decline in crop yield following market liberalization and removal of subsidies that permit fertilizer use. The average fertilizer use intensity in 2002/2003 cropping season was only 9 kg ha^{-1} of harvested land. Africa is rich with indigenous phosphate rock deposits, which should provide an incentive for direct application or local chemical treatment at low cost, to improve the solubility of the low reactive phosphate rocks. Since fertilizers cost about 2-4 times more at the farm gate in Africa than in most continents of the world, the availability of indigenous phosphate rock deposit provides a robust natural resources management to alleviating P-deficiency in soils of Africa.

Sensible water management approach

The expansion of any sector's water use will require better management to save fresh water, improved water harvesting techniques, desalinisation of sea water and recycling of waste water (Spore, 2001). Water-saving techniques for rain-fed crops should aim essentially to: (i) increase water infiltration into the soil and decrease runoff, (ii) facilitate plants to use water stored in the soil and, (iii) promote better conservation of water infiltrated in the soil. The technique of tied ridge consists of making small dykes at regular intervals in the furrow. Runoff is almost eliminated and all the water is captured in the furrows for infiltration. The date of tying the ridges is very important. It must coincide with when rains are heaviest and when the growing crop needs water most. In a trial with upland cotton, tying ridges at 6 week after planting give the optimum seed cotton yield in all the years (Table 2).

Rock bunds are another technique to combat runoff and erosion. Rock bunds are of three types: (i) a 3-stone system, (ii) a single row of stone, and (iii) a stock of stones embedded in the earth. Under this conservation practices, the runoff coefficient had reduced from 50% to 10% (Ouattara et al., 1999).

Appendix

The potential for irrigation development has not been effectively tapped in Africa. Only 12.6 m ha or 3.7% of the surface area of Africa is currently under irrigation. The proper management of irrigation is the key to sound water management. In addition to improving the efficiency of existing irrigated lands, there is also some potential of expanding irrigated land area. Rather than large scale irrigation schemes (e.g. large dams), emphasis needs to be on small-scale irrigation projects involving small farm units. Appropriate small scale irrigation schemes may involve use of ground water, runoff storage and alternate furrow irrigation. Alternate furrow irrigation means supplying every other furrow with irrigation water. Experiments indicate that alternative furrow irrigation is an innovative irrigation water management technique. The crops grown will use about half as much water as the traditional system and loses two-third less in runoff, resulting in water savings and enhanced productivity.

Integrated approach

Efforts for dealing with efficient natural resource management for increased agricultural sustainability must benefit from integration of the various corrective strategies with a more holistic approach to land management, e.g. conjunctive use of organic and inorganic fertilizer and parallel use of soil water and nutrient management systems. Wider adoption of integrated land management, its further development and improved management of input use, provide the main technological way to meet the challenge of required increases in agricultural productivity. Sanchez (2001), proposed an integrated approach consisting of bringing nitrogen from the air (biological nitrogen fixation) and phosphorus from indigenous phosphate rock deposits (inorganic fertilizer), together with biomass transfers of a nutrient-accumulating hedge species as a source of rapidly available nutrients and carbon (agroforestry), as an example of integrated resource management in sustainable agricultural productivity.

References

Adeboye, M. K. A., Iwuafor, R. N. O. and J. O. Agbenin. 2006. The effects of crop rotation and nitrogen fertilization on soil chemical and microbial properties in a Guinea savannah Alfisol of Nigeria. *Plant and Soil* (in press).

Adeoye, K. B., 1986. Long-term effects of dung on the physical properties of a loamy savannah soil of northern Nigeria. Pp 81-83. In: Proceedings of Soil Science Society of Nigeria, Makurdi, Nigeria.

Agnew, C. T., 1986. Water availability and the development of rain-fed agriculture in South-West Niger, West Africa. *Trans. Inst. Br. Geogr. N.S.* 7: 419-457.

Bationo, A., Mokwunye, U., Vlek, P. L. G., Koala S. and B. I. Shapiro. 2003. Soil fertility management for sustainable land use in the West African Sudano-Sahelian zone. Pp 247-292. In: M. P. Gichuru *et al.* (eds.). Soil Fertility Management in Africa: A Regional Perspective. Academy Science Publishers.

Craswell, K. T. and R. D. B. Lefrog. 2001. The role and function of organic matter in tropical soil. *Nutrient Cycling in Agroecosystems* 61: 7-18.

FAO (Food and Agricultural Organization of the United Nations), 1999. Agricultural policies for sustainable management and use of natural resources in Africa. FAO/IITA Joint Publication, 81 pp.

ICRISAT (International Crops Research Institute for the Semi-Arid Tropics), 1991. ICRISAT West African Programs, Annual Report, 84 pp.

INRA (Institute for Natural Resources in Africa), 1995. Land restoration and integrated watershed management in sub-Saharan Africa: A project proposal. INRA-UNU, Accra, Ghana.

Kang, B. T., Hauser, S., Vanlauwe, B., Sanginga N and A. N. Atta-Krah, 1995. Alley farming research on high base status soils. Pp 25-39. In: Alley Farming Research and Development. B. T. Kang et al. (eds.). Proceedings of an International Conference on Alley Farming. Intee Printers Ltd, Ibadan.

Kimani, S. K., Nandwa, S. N., Mugendi, D. N., Obanyi, S. N., Ojiem J., Murwira, H. K. and A. Bationo, 2003. Principles of integrated soil fertility management. Pp 51-72. In: M.P. Gichuru et al. (eds.). Soil Fertility Management in Africa: A regional Perspective. Academy Science Publishers.

Lal, R., 1995. Erosion-crop productivity relationships for soils of Africa. Soil Sei. Soc. Am. J., 59: 661-667.

Lal, R. and B. Okigbo, 1990. Assessment of soil diction in the southern states of Nigeria. The World Bank Sector Policy and Research Staff Environment Working Paper No. 39.

Lal, R., Hassan H. M. and J. Dumanski, 1998. Desertification control to sequester carbon and mitigate the greenhouse effect. Pp 83-149. In: N. J. Rosenberg, R. C. Izaurralde and K. L. Malone (eds.). Carbon sequestration in soils: Science, monitoring and beyond.

Lawson, T. C. and M. V. K. Sivakumar, 1989. Climatic constraints to crop production and fertilizer use. Fertilizer Research 29: 9-21. Nandwa, S. M. 2003. Perspectives on soil fertility in Africa. Pp 1-50. In: M. P. Gichuru et al. (eds). Soil Fertility Management in Africa: A Regional Perspective, Academy Science Publishers.

Obaid, T. A., 2004. Health and the links to nutrition: Maternal health is key. In: Nutrition and the Millennium Development Goals. SCN News (A Periodic Review of Development in International Nutrition): 15-18.

Odunze, A. C., Tarawali, S. A., de Haan, N. C., Iwuafor, E. N. O., Katung, P. D., Akoueguon, G. E., Amadji, A. F., Schultze-Kraft, R., Atala, T. K., Ahmed, B., Adamu, A., Babalola, A. O., Ogunwole,J. O. A., Alimi, A., Ewansiha, S. U. and S. A. Adediran, 2004. Grain Legumes for Soil Productivity Improvement in the Northern Guinea Savannah of Nigeria. Food, Agriculture and Environment, 2(2): 218-226.

Ogunwole, J. O., 2004. Effects of fertilizer and time of ridge-tie on yield and fibre quality of late sown cotton in the dry savannah zone of Nigeria. *J. Sustainable Agriculture,* 24(3): 97-107.

Okigbo, B. N., 1996. Crops and cropping systems in sub-Saharan Africa. Pp 589-617. In: V. L. Chopra, R. B. Singh and A. Varma (eds.). Crop Production and Sustainability: Shaping the Future. Proceedings of the 2nd International. Crop Science Congress. Oxford and IBH Publishing Co. PVT Ltd. New Delhi, India.

Ouattara, B., Hien V. and F. Lompo, 1999. Development of water management technologies for rain-fed crops in Burkina Faso. Pp 265-281. In: N. van Duivenbooden *et al.* (eds.). Efficient soil water use: the key to sustainable crop production in the dry areas of West Asia, and North, and Sub-Saharan Africa. Proceedings of the 1998 (Niger) and 1999 (Jordan) Workshops of the Optimising Soil Water Use (OSWU) Consortium, Aleppo, Syria, ICARDA, and Patancheru, India, ICRISAT.

Pereira, L. S., 2005. Water and Agriculture: Facing water scarcity and environmental challenges. Agricultural Engineering International. The CIGR Journal of Scientific Research and Development. Invited Overview Paper. Vol. VII 26 pp.

Rashid, A., 2006. The future of soil science in less developed countries. Pp 119-121. In: The Future of Soil Science, International Union of Soil Sciences.

Sanchez, P. A., 2001. Multifunctional agriculture in the tropics: Overcoming hunger, poverty and environmental degradation. Pp xvii-xxxiv. In: Agricultural Technology Research and Sustainable Development in Developing Region. Proceedings of the 7th JIRCAS International Symposium, November 1-2, 2000. Tsukuba, Japan.

Sivakumar, M. V. K., 1997. Climate variability and food production. Scientific lecture delivered at the 49th Session of the Executive Council of the World Meteorological Organization. June, 10-20, Geneva, Switzerland.

Sivakumar, M. V. K. and J. S. Wallace. 1991. Soil water balance in the Sudano-Sahelian zone: Need, relevance, and objectives of the workshop. Pp 3-10. In: M. V. K. Sivakumar, J. S. Wallace, C. Renard and C. Giroux (eds.). Soil Water Balance in the Sudano-Sahelian Zone. IAHS Publication No. 199.

Spore, 2001. Management of Water Resources: Ways with Water. ACTA publication, Number 95: 4-5.

Spore, 2003. Ageing and Agriculture: A hard rock of age. ACTA publication, Number 102: 1-2.

Uyovbisere, E. O. and K. Elerno, 2002. Effect of tree foliage of locust bean *(Parkia biglobosa)* and neem *(Azadirachta indica)* on soil fertility and productivity of maize in a savannah alfisol. Nutrient Cycling in Agroecosystems, 62: 115-122.

Vanlauwe, B., 2004. Integrated Soil Fertility Management Research at TSBF: The Framework, the Principles, and their Application. Pp 25-42. In: A Bationo (eds.). Managing Nutrient Cycles to Sustain Soil Fertility in Sub-Saharan Africa. Academy Science Publishers.

Appendix

Time of ridge-tie	Seed cotton yield (kg ha^{-1})		
	1999	2000	2001
Open ridge (No-tie)	444.5b	576.8b	796.7a
Tying @ 4 W AP*	1137.5a	854.3a	889.6a
Tying@ 6 WAP	1413.9a	910.7a	948.3a
Tying@ 8 WAP	346.2b	964.4a	965.5a
SE±	123.1	87.4	94.9

Table 2. Yield of seed cotton (kg ha) as influenced by time of ridge-tie Adapted from Ogunwole (2004), * WAP = Weeks after planting.

SOIL SCIENCE AND SMALLHOLDER AGRICULTURE IN SUB-SAHARAN AFRICA

Emmanuel Uzoma Onweremadu

Abstract

Smallholder farmers in sub-Saharan Africa are operating in a highly variable and complex environment with soils varying considerably over short distances in soil fertility levels. This situation is aggravated by the spate of several forms of degradation ravaging the region. All these make blanket treatment of soils ineffective. Socio-cultural peculiarities of the area make most alien technologies infeasible and non-efficacious. It becomes a matter of necessity to evolve technologies which are improvements on the existing ones with farmers being part of the evolution for sustainable soil management. Further enhancement of soil resource management can be attained if Decision Support Tools (DSTs) are used in research efforts in the sub-Saharan Africa.

Keywords: *Agriculture, low-input farmer, soil, sustainability, tropics.*

Introduction

Africa's basic industry is agriculture, providing about 35% of the gross national product (GNP), 40% of exports and 70% of employment. Given its size, agriculture should be the engine of economic growth. Nevertheless, living conditions are desperately pour, about 240 million citizens live on less than US $ 1 a day, primarily as smallholder farmers (Struif Bontkes and Wopereis, 2003). Yet, sub-Saharan Africa has the highest fertility rate in the world, thus faces increasing population pressure on its natural resources (Franzel *et al.,* 2004). Amalu (2002) reported that about 168 million inhabitants of Africa are chronically malnourished. This is despite the fact that more than two-thirds of Africans are farmers (Henao and Baanante, 2001). African farming is essentially a low-input low-output system (Badiane and Delgado, 1995) and this is associated with soil degradation, particularly in the absence of appropriate conservation practices (Scherr and Yadav, 1996). The landless are clearing forests, eking out a living on poor quality soils (Barnes, 1990) due to insufficient prime agricultural lands. He reported that forest consumption is about 3 million hectares per annum and forests are cleared by the traditional slash-and-burn method (Ngugi, 1996) which releases CO_2 into the atmosphere, thereby contributing to global warming. In addition to this, harsh climate promotes soil infertility creating intense pressure on land even at relatively low population densities.

Three basic concerns in sub-Saharan Africa are population growth, agricultural performance and environmental degradation. As population increased from an average of 2.5% in the 1960s to more than 3% in the 1990s, demand for food heightens and land is increasingly permanently cropped without adequate inputs since farmers lack purchasing power. Cultivated land per capital has fallen by 40% since 1965 from 0.5 to about 0.3 hectare per person with increased demographic pressure. These result in soil degradation. There is therefore a causal nexus between land, population, poverty and soil degradation.

More than four decades of research and development work in Africa have not resulted in the 3-5% annual increase in agricultural growth (Badiane and Delgado, 1995) that is necessary for most African countries to ensure sustainability of agriculture and the promise of food security in the nearest future. Sluggish or zero growth is likely because of the cumulative effect of many factors but with strong bearings on soil resource and its productivity.

The Problem

Tremendous short-distance intra- and inter-pedon variability in soil fertility exists in African soils, making blanket recommendation ineffective. They have suffered a lot of stresses leading to various species of degradation, such as nutrient depletion, low nutrient holding capacity, low water holding capacity, extreme soil temperatures, aluminium toxicity, low structural stability, root restrictions due to compaction and high shrink-swell properties. The study by Oldeman *et al.* (1991) indicates that soils on about 5 million hectares of land in Africa are degraded to a point where their original biotic functions have been fully destroyed and resilience reduced to such a level that rehabilitation to make them productive may be economically prohibitive. Up to 400 million tonnes of topsoil are washed away annually into South Africa rivers, dams and the sea, and the available prime lands are threatened by soil erosion, chemical pollution and land uses, such as residential development and mining (Nduli, 2000).

Farming without adequate soil fertility regenerative inputs eats away the continent's soil budget. When a farmer notices a decline in soil fertility on the plot of land he is farming, he looks for another land whether it is arable or not. Farming is practiced on any kind of slope. Data on nutrient balances over the past 30 years suggest that African soils have sustained annual net losses of nitrogen, phosphorus, and potassium on the order of 22, 2.5 and 15 kilograms per hectare, respectively. The nutrient balances for Southern Mali as a whole are negative, especially for nitrogen and potassium (Stoorvogel and Smaling, 1990).

An alarming concept that African soils are being steadily depleted of nutrients due to farming without appropriate inputs has gained wide credence in the scientific community (Brinkman, 1990; Lai, 1994; Bationo *et al.*, 1998; Gruhn *et al.*, 2000; Eswaran *et al.*, 2001). Soils show extreme levels of chemical and physical conditions, such as very high or very low pH (Lai, 1994) as well as inability to recover from mismanagement (Brinkman, 1990).

Farmers are aware of soil degradation and associated poor yield. Despite this, most non-African technological packages are not adopted by African farmers since most of them do not fit into African farmer rationality. Non adoption could be for pragmatic reasons, where the technology being promoted was not suited to the new environment or social context (Frank and Chamala, 1992). Again, most of the alien technologies are complex and non-divisible. Farmers act quite rationally by preferring to adopt less complex innovations over more complex ones and by not adopting complex practices at all. Failure of foreign technologies stems from their inadequate verification, validation and certification for field use in Africa (Greenland *et al.*, 1994).

Now, old strategies for coping with new pressures on soil resource base are becoming increasingly infeasible. Fallow length for fertility restoration has shortened and sometimes non-existent. Fertilizer subsidies are withdrawn during the structural adjustment liberalizations of the 1990s. Rural credit systems have collapsed and this has rendered reliance on chemical fertilizers increasingly unprofitable and infeasible to farmers. The cultural background of the study area does not empower women economically while they remain key players in African farming. The menace of Human Immunodeficiency Virus (HIV)/Acquired Immune Deficiency Syndrome (AIDS) has drastically reduced asset building capacity of farmers.

Soil data are scanty and are not translated into problem-solving technology (Lai and Ragland, 2003), and this could be part of the reason why available soil data are rarely used. There is insufficient resource information on land type and land use in land types. Ecotypes in land types are not described and mapped. Conservation status of soil resource is not determined. National and regional degradation manual is non-existent in most African countries, thus indicators and classes are not known. The geographical information systems (GIS) are not functioning and user manual for GIS scientists are lacking. There is no database structure and sufficient basic data in the continent's GIS.

Based on these and more the purpose of this contribution is to stress the relevance of soil and soil-related information in the sustainable management of soil resource in the sub-Saharan Africa. The writer considers soil as a polyvalent natural resource intervening in a large number of other resources and human activities. Soil has agronomic, engineering, sanitary, aesthetic and recreational utilities (Zinck, 1990). Consequently, soil degradation has negative impact on food security, air and water quality, landscape beauty, morbidity and general health of citizens of sub-Saharan Africa.

Attributes of Soil Resource

More often than not, soil is considered as a resource unlimited in time and space. But soil has three main attributes, namely scarcity, vulnerability and low resilience. Although the soil mantle forms a continuum with local interruptions, good soils for agriculture are scarce. Prime agricultural lands represent only a small proportion of the worldwide soil budget. Soils with few stresses occupy only 0.4% of the total land area in Africa (Reich et al., 2004). They reported that 43% of Africa is vulnerable to desertification.

Soils are susceptible to deterioration if mal-handled. Even the solid high productive loess-derived soils show increasing signs of exhaustion and physical fragility. In the tropics, many soils are in a fragile equilibrium even before they are put under agricultural use.

African society often considers soil as a renewable resource. Soil loss tolerance figures for example are frequently established as if a truncated or eroded soil would recover in a human time scale. In fact, it takes centuries to millennia to form a sizeable thickness of soil cover appropriate for plant growth (Zinck, 1990).

Methods

Soil Science (Soil Survey) captures data useful for soil and soil-related decisions from four main resources, namely field observations and determinations (ground truth), remote sensing techniques, laboratory analyses, and expert opinions and experimentation results. All these methods are currently enhanced using the geographic information systems (GIS) technologies, especially in presenting outputs (memoir and maps) in customized forms.

Visual interpretation of conventional aerial photographs remains the most comprehensive technique of data extraction for soil surveyors at medium and large scales. It is comprehensive since it allows for both a relatively accurate delimitation of soil map units and a satisfactory prediction of map unit features, external as well as internal. Assisted by a stereoscope, visual interpretation of Radio Detection and Ranging (RADAR) and spectral images is made possible at smaller scales. The ground penetrating RADAR (GPR) devices can provide precise images of complete soil sequences and is capable of penetrating 1 meter depth in compacted clayey soils to 25 meters in sandy soils (Doolittle, 1987). High spectral and spatial resolutions as seen in digital processing improves soil cartography, especially regarding the boundary precision of soil map unit delineations. High temporal resolutions help monitor ever-changing surface features and processes, such as urban fringe dynamics, crop rotations, erosion, urbanization, salinisation, desertification, flooding, etc. Video image analysis (VIA) of large-scale vertical air-photos permits identification of detailed surface features, in particular drainage conditions and monitoring of changes at a large scale say 1:5000 (Harrison *et al.*, 1987). However, cost of remote sensing techniques makes them rarely used at the farm level but government support can make it affordable.

Field observations and determinations are simple and depend heavily on experience. Field observations are often used to check results of aerial surveys. This is referred to as ground truth, which in itself is indispensable in establishing the composition of the soil map units in terms of taxa components present, including their respective geographical extent. Observation faculty is the most important, intangible tool a soil surveyor uses in the field. The low-income farmers in Africa use this technique liberally in identifying suitability of farmlands for cropping. Modern surveying has improved data acquisition through this technique by using penetrometers, small portable laboratories microcomputers, GPR, electromagnetic induction (EM), Doppler satellite and Global Positioning System (GPS) receiver.

Laboratory analyses support field studies and observations. Routine and special analyses are carried out. Often selective sampling is done to reduce laboratory costs. When supported by high-quality field description of soil, selective sampling can indeed be used to minimize cost. However, composite sampling is used to estimate short-range variability. It has been recognized as having strong influence on statistical dispersion of laboratory results.

Expert opinion and experimentation results are important in data acquisition. It is common to use experimental plot and farmer data to cross-check and calibrate land evaluation results, especially in yield predictions. In contrast, a domain which has hardly been explored is the elaboration of soil maps through systematisation of farmer and peasant expertise. Past civilizations like the Aztec ethnopedology, because of their intimate contact with and dependency on soil resource, developed a rich vocabulary on relevant soil properties and soil types. Today's village farmers represent excellent sources of soil information for knowledge-based soil map elaboration. Again, engineering expertise in addition to records on soil behaviour and failure cases is basic for non-agricultural soil interpretations.

To enhance incremental land use planning, geographic information system is opined since the technology is capable of collecting, storing, retrieving, transforming and displaying spatial information from the real world (Burrough and McDonnell, 1998). According to Brady and Well (1999), a GIS has five steps: data acquisition, pre-processing, data management, manipulation and analysis of generated data. Outputs of GIS are in customized and understandable forms (Kufoniyi, 2000).

Soil survey uses questionnaires, rapid appraisal techniques, on-and off-farm demonstrations, workshops with farmers, farmers' days, agricultural shows and participant observation of farmers' practices to get information on socio-economic data. These techniques enhance the reliability and predictability of the earlier popular methodologies of soil survey. Soil fertility management practices were identified through interviews, transect walks, soil and nutrient flow maps and developed by farmers in Kenya and Uganda (De Jager *et al., 2003).*

Soil management options

The spectrum of usage of soil information has widened and it is needed by farmers, engineers, environmentalists, planners, health experts, miners, foresters and other natural resources managers (Lee and Krall, 2004). In the light of the above, the productive capacity of African soils is paramount. Larson and Pierce (1991) view soil quality as the capacity of soils to function within its ecosystem boundaries and interact positively with the environment external to that ecosystem. They linked soil quality to four sustainability objectives of Lourance (1990) which encompasses agronomic, ecological, micro and macro-economic sub-areas.

Given that the production levels of smallholder farmers are often far below what would be possible under improved circumstances, it is important to consider the diversity and dynamics of farmer reality (Scones, 2001). The diversity of Africa farmer reality implies that solutions need to be evolutionary, site-specific and with much emphasis on farmer experimentation, participatory learning and building of partnerships between soil fertility management stakeholders (farmers, credit providers, input dealers, researchers, extension workers and government) at village regional and national levels.

Soil fertility management needs to play a central role in interventions aimed at improving agricultural productivity in a sustainable manner. Options should rely on soil nutrient-supplying capacity, available soil amendments and judicious use of fertilizers (organic and inorganic) to achieve a balanced nutrient management system.

Soil survey tool is the first field step towards sustainable soil resource use. The tool provides a quantitative basis for evaluating different land use options and impacts of technology, furnishes, parameters for quantifying ecosystem interactions, evaluates statuses and impacts of soil conditions over given periods of time, provides a basis for targeting conservation programmes, enhances environmental assessment and provides a basis for identifying tension (risk) zones.

Farmers and other soil users should participate in soil mapping on this, a soil typology, is established by farmers through field visits, digging of pits (minipedons) and other field experiences. Sub-regional air photo mosaics can be given to such survey teams but assisted by a soil surveyor. The communal nature of African socio-cultural background can be a driving force in achieving successes. At the end of the field experience, farmer maps will produce fertility classes based on spatial variability, thus fertility capability classification (FCC). These maps can then be scanned, digitised and fed into a GIS environment created by appropriate software. Products of GIS query are taken as farmers' properties which serve as a guide in future land use(s). Farmers now realize that soils are vulnerable and that such soils span several properties. They also know that measures needed to prevent further soil erosion and solination require conservation practices, such as alignment of fences and vegetation belts.

Farmer involvement is necessary in soil resource management. Many of the innovations underway have been developed by farmers themselves. Because of the variable soils and farmer conditions and preferences, a range of technologies and alternatives remain necessary in most locations. Farmers should preferably be confronted with several possible options. In the case of Mono province in the Republic of Benin, four options were proposed to address severe soil fertility problems. Two were for completely exhausted fields (planting fallow with *Acacia auriculiformis* and relay planting of *Mucuna pruriens,* two for less exhausted fields (alley cropping and a short fallow with *Cajanus cajan,* interplanted with maize).

Sub-Saharan Africa is endowed with a panorama of vegetal types and this underscores high potentials for organically-based fertility technologies. With a rapidly increasing world population and shrinking prime farmlands, higher amounts of organic matter pool transfers will be inevitable, so that efforts have to be made to optimise the utilization of the annual nutrient recycling pool. The point of zero charge (PZC) of humic substances is as low as pH 3.0 to 3.5, which underlines the importance of humic substance as an exchanger for cations, especially in oxidic low-activity clay soils, because it can be deduced that in these soils, the oxidic soil minerals, the amorphous silica and the whole clay fraction have a PZC above the soil pH.

Soil fertility restoration technologies should mimic the natural vegetal cover. Improved fallows stand out in this regard. The variable nature of the tropical vegetation must be considered. Because most shrubs and trees have long gestation period, attempts should be made to genetically engineer them for rapid growth. Adopting improved fallows using fast-growing and soil fertility enhancing crops is ideal. This strategy should be practiced in areas where land is not a limiting factor. Two-and three-year *Sesbania sesban-based* fallows have proved highly effective in soil fertility restoration in Malawi and Zambia (CTA, 2002). Maize grain yields following a 3-year *Sesbania sesban* fallow without fertilizer in Chipata, Zambia, were 2.27, 5.59 and 6.02 tha-^{1} after 1, 2 and 3 years fallow, respectively compared with the control plots with 1.6, 1.2 and 1.8 tha-^{1} after 1, 2 and 3 years of continuous cropping (Kwesiga and Chisumpa, 1992).

Inter-planting maize and *Sesbania* seedling concurrently but maintaining the desired maize population is a good soil fertility restoration technology. After harvesting the maize, the *Sesbania* trees are left to grow during the ensuing dry season, utilizing residual moisture. The trees are cut down before the following planting season and leafy residues are incorporated into the soil. This relay cropping technology is best suited to densely populated areas where farm sizes are too small to accommodate normal rotational fallows. *Sesbania* trees supply fuel-wood which is of added advantage in areas experiencing shortages (Maghembe *et al., 1997)*.

It is laborious planting trees every 2-3 years in improved fallows as *Sesbania* sesban-based types. However, mixed intercropping of maize and nitrogen fixing *Gliricidia* is an alternative technology to the above. It is adapted to repeat cutting back without dying. In the Shire highlands in Malawi, as many as 2-4 pruning are obtained each year, giving 2.7 tha-^{1} biomass and Ikerra *et al.* (1999) reported that 60 to 120kg/ha/yr nitrogen equivalent was added into the soil. Yields from third year onwards are markedly increased to an average of 1800-2500kg/ha (Bohringer and Akinnifesi, 2001). Mixed farming is advocated for mutualistic co-existence of livestock and crops. Animal droppings are often carried to the farm near the homestead to enrich soils especially for high-value crops (Scoones, 2001). Cropped fields distant from homestead rarely receive animal manures hence are often highly degraded. The use of compost pits established in such locations can assist in soil fertility regeneration and improvement in such distant farmlands.

Biomass transfer is another viable technology. It involves growing nitrogen-rich biomass *in situ*, cutting and transporting it, and applying it to plots where high-value crops are grown. The biomass plots are linked to organic fertilizer banks. It is popular in Chipata, Zambia, for growing vegetables. *Tephrosia, Leucaena* and *Sesbania* are good sources for biomass technique.

Rampant deforestation, overgrazing and use of steep slopes for cultivation have resulted to soil erosion. Improved fallows and use of vetiver grass strips on contour banks will assist in minimizing accelerated erosion. Agro forestry, growing high-value crops, rotational woodlots and fodder banks are good for soil and water conservation. Use of planting basins known as *Zai* in Burkina Faso has been proved successful as a conservation farming technology.

Implementation of these technological options should consider the psychology of localities. Several countries, including Zambia, Zimbabwe, Kenya and Uganda, disseminate information effectively through farmer groups, farmer-to-farmer learning strategies, farmer visits, field days and demonstrations. This may be different in other locations.

Furthermore, Decision Support Tools (DSTs) in research and extension are rarely used in sub-Saharan agriculture. An important reason for this is that many research projects focus on the introduction of one single tool, where as a systems approach is clearly needed to cover the diverse and sometimes contrasting demands of the farmer and agro-ecology. The use of DSTs may allow cost and time savings and improve the quality of decision-making by smallholder farmers. In 1999, the Africa Division of International Centre for Soil Fertility and Agricultural Development (IFOC) started to develop, evaluate, and promote DSTs through BA Client-Oriented Systems Toolbox for Technology Transfer Related to Soil Fertility Improvement and Sustainable Agriculture in West Africa (COSTBOX).

Conclusion

Agriculture remains the key to Africa's economic future. With approximately 200 million people chronically hungry and malnourished, it becomes necessary to improve on the current technologies of soil fertility regeneration and improvement, since they are not keeping pace with the food demands of the day. In doing this, it should be noted that soil is a polyvalent natural resource influencing other natural resources, such as water, air, forest and wild animals; hence its deterioration affects these resources.

Technological options for sustainable management must stem from existing indigenous ones and for efficacy, farmers' participation and rationality must be considered in the evolution of traditionally-based improved practices.

Table 1. Extent of human induced, nutrient-related soil degradation in selected regions (million hectares)

Region	Light Degradation	Moderate degradation	Severe degradation
Africa	20.4	18.8	6.6
Asia	4.6	9.0	1.0
South America	24.5	31.1	12.6

(Source: Oldeman, MakkelingiSombroek, 1992)

Table 2: Human-induced causes of soil degradation (percent)

Region	Deforestation	Overpopulation	Overgrazing	Agricultural activities	Industrial Activities
Africa	13.6	12.8	49.2	24.5	...
Asia	13.9	6.2	26.4	27.3	0.1
South America	41.0	4.9	27.9	26.2	...
World	29.5	6.8	34.5	28.1	1.2

(Source: Oldeman, 1992)

Table 3: Annual growth rates in fertilizer use per hectare, 1960s-1990s (percent)

Period	West Africa	East Africa	Southern Africa	Sub-Saharan Africa	Developing world	World
1960s	14.2	10.2	5.7	11.	15.4	9.4
1970s	17.6	1.3	10.4	4.8	9.9	5.2
1980s	4.0	1.5	-3.9	1.9	4.6	2.2
1990s*	8.8	0.7	-3.4	3.2	3.1	0.5
1960s-1990s	9.7	4.2	3.4	5.4	8.3	3.7

Source: Computed by the author using data from FAO 1998 and 1999 * 1990s = 1990-1996.

92

References

Amalu, U. C., 2002. Food security: Sustainable food production in sub-Saharan Africa. *Outlook on Agriculture,* 31 (3): 177-185.

Badiane, O. and C. L.. Oelgado, 1995. A 2020 vision for food, agriculture and the environment in sub-Saharan Africa. Food, Agriculture and Environment Discussion Paper 4, International Food Policy Research Institute, Washington. D.C., 56 pp.

Barnes, O. F., 1990. Population growth, wood fuels, and resources problems in sub-Saharan Africa. Industry and Energy Department. Working Paper No. 26. Washington, D.C. World Bank.

Bationo, A., Lompo, F. and S. Koala, 1998. Research on nutrient flows and balances in West Africa State-of-the art. *Agriculture, Ecosystems and Environment,* 71: 19-35.

Bohringer, A. and F. Akinnifesi, 2001. The way ahead for the domestication and use of indigenous fruit tress from the Miombo in Southern Africa. ICRAF, Makoka, Malawi.

Brady, N. C. and R. R. Well, 1999. *The nature and properties of soils* 12th ed. Prentice Hall, New Jersey.

Brinkman, R., 1990. Resilience against climate change. In: Scharpenseel, H. W., Shoemaker, M. and A. Ayoub (eds.) *Soil on a warmer earth.* Amsterdam: Elsevier. Pp. 51-60.

Burrough, P. A. and R. A. McOonnell, 1998. *Principles of Geographie Information Systems,* Oxford University Press, 208 pp.

CTA (Technical Centre for Agricultural and Rural Cooperation), 2002. Agroforestry in Malawi and Zambia. Summary report of a CTAIMAFE study visit. CTA, Wageningen, The Netherlands 32 pp.

De Jager, A., Onduru, D. and C. Walaga, 2003. Using NUTMON to evaluate conventional and low external input farming practices in Kenya and Uganda In: Struif Bontkes T. E. and M. C. S. Wopereis, (eds). *Decision support tools for smallholder agriculture in sub-Saharan Africa: A practical guide* CTA. Wageningen, The Netherlands. pp 10-53.

Doolittle, J. A., 1987. Using ground penetrating radar to increase the quality and efficiency of soil surveys SSSA spec Publ. No. 20, Madison, pp. 11-32.

Eswaran, H , Lai, R, and P. F. Reich, 2001. Land degradation: An overview. In: Bridges, E M., Hannam, I. O., Oldeman, L. R., Penning de Vries, F. W. T., Scherr, S. J., and S. Sombatpanit, (eds). *Response to land degradation.* Enfield, New Hampshire, USA; Science Publishers, pp 20-35.

Franzel, S., Place, F., Reij, C. and G. Tembo, 2004. Strategies for sustainable natural resources management. In: S. Haggblade, (ed.). *Building on* successes *in African agriculture*. International Food Policy Research Institute, 2020 Vision for Food, Agriculture and the Environment, Focus 12 Brief 8 of 10.

Greenland, D. J., Bowen, G., Eswaran, H., Rhoades, R. and C. Valentin, 1994. Soil, water and nutrient management research: A new agenda. IBSRAM, Bangkok, Thailand. 72 pp.

Gruhn, P., Goletti, F. and M. Yudelman, 2000. Integrated nutrient manage-ment, soil fertility, and sustainable agriculture: Current issues and challenges. International Food Policy Research Institute, Washington, D. C. Food, Agriculture and Environmental Discussion Paper 32. 31 pp.

Harrison, W. O., John, M. E. and P. F. Biggam, 1987. Video image analysis of large-scale vertical aerial photograph to facilitate soil mapping. SSSA Spec. Publ. No. 20, Madison, pp 1-9.

Henao, J., and C. Baanante, 2001. Nutrient depletion in the agricultural soils of Africa. In: Pinstrup-Andersen and R. Pandya-Lorch (eds) *The unfinished agenda: Perspectives on overcoming hunger, poverty, and environmental degradation*. International Food Policy Research Institute, Washington, D. C. pp. 159-163.

Ikerra, S T, Maghembe, J. A., Smithson, P. C., and R. J. Buresely, 1999. Soil nitrogen dynamics and relationships with maize yields in a Gliricidia-maize intercrop in Malawi. *Plant and Soil,* 211: 155-164.

Krall, L. and C. E. Lee (2004). Keeping the link: Soil survey inventory and technical service. *Soil Survey Horizon,* 45 (1): 35-36.

Kufoniyi, O., 2000. Basic concepts in geographic information systems (GIS). In: Ezeigbo, C. U. (ed) *Principles and practices of GIS.* Panuf Press, Lagos. Pp 1-15.

Kwesiga, F, and S. M. Chisumpa, 1992. Multipurpose trees for the Eastern Province of Zambia: An ethnobotanical survey of their use in the farming systems. AFRENA Report No. 49. ICRAF, Nairobi, Kenya.

Lai, R., 1994. Methods and guidelines for assessing sustainable use of soil and water resources in the tropics. SMSS Technical Monograph No. 21. Washington OC: NRCS 78 pp.

Lai, R. and J. Ragland, 1993. Agricultural sustainability in the tropics: Technologies for sustainable agriculture in the tropics, ASA Spec. Publ. 56 1-6.

Larson, W. E. and F. J. Pierce, 1991. Conservation and enhancement of soil quality In: Dumanski, J., Pushparajah, E., Latham, M. and R. Myers (eds). *Evaluation for sustainable land management in the developing world.* International Board for Soil Research and Management, 2: 175-204.

Lourance, R., 1990. Research approaches for ecological sustainability. J. *Soil and Water Conserv., 45:51-54.*

Maghembe, J. A., Chirwa, P. W., Kooi, G. and S. Ikerra, 1997. SADC-ICRAF. Annual Report for 1997. ICRAF, Makoka, Malawi.

Nduli, N. J., 2000. Information and institutional systems required for natural resources policy formation in South Africa. In: *Proceedings of a CTA Workshop on Information Support for natural resource management policy.* Pp 145-158.

Ngugi, D., 1996. Agroforestry for sustainable rural development in the Zambezi basin. Project Management Plan. ICRAF, Nairobi, Kenya.

Oldeman, L. R, Hakkeling, R. T. A. and W. G. Sombroek, 1991. World map of the status of human-induced soil degradation: An explanatory note. ISRIC-UNEP Report. The Netherlands. 43 pp.

Oldeman, L. R., 1992. Global extent of soil degradation. In: ISRIC Bi-annual Report 1991-1992. Wageningen, The Netherlands: International Soil Reference and Information Centre.

Oldeman, L. R., Makkeling, R.T. A. and G. W. Sombroek, 1992. World map of the status of human-induced soil degradation: An explanatory note. 2nd ed. Wageningen, The Netherlands: International Soil Reference and Information Centre.

Reich, P. F., Numbem, S. T., Almaraz, R. A. and H. Eswaran, 2001. Land resource stresses and desertification in Africa. In: Bridges, E. M, Hannam, I. O, Oldeman, L. R., Pening de Vires, F. W. T, Scherr, S. J. and S. Sompatpanit, (eds). *Responses to land degradation* Proc. 2nd International Conference on Land Degradation and Desertification, Khon Kaen, Thailand.

Scherr, S. J. and S. Yadav, 1996. Land degradation in the developing world: Implications for food, agriculture and the environment to 2020. Food, Agriculture and Environment Discussion Paper 14. International Food Polcy Research Institute, Washington, DC. 36 pp.

Scoones, I., 2001. Transforming soils: The dynamics of soil-fertility management in Africa. In Scoones, I. (ed). *Oynamics and diversity: Soil fertility and farming livelihoods in Africa* London, Earthcan. Pp 1-44.

Stoorvogel, J. J. and E. M. A. Smaling, 1990. Assessment of soil nutrient depletion in sub-Saharan Africa, 1983-2000, Vol. I-IV, Wageningen, The Netherlands The Winand Staring Centre.

Struif Bontkes, T. E. and M. C. S. Wopereis, 2003. Opportunities for the use of decision support tools for smallholder agriculture in sub-Saharan Africa. In: Struif Bontkes, T. E. and M. C. S. Wopereis (eds). *Decision support tools for smallholder agriculture in sub-Saharan Africa: A practical guide.* CT A The Netherlands. 123 pp.

Versteeg, N. M. and V. Koudokpon, 1993. Participative farmer testing of four low external input technologies to address soil fertility decline in Mono Province, Benin. *Agricultural System,* 42: 265-276.

Zinck, J. A., 1990. Soil survey epistemology of a vital discipline. *ITC Journal,* *4:335-350.*

APPLICATION OF ORGANIC FARMING FOR SUSTAINABLE HORTICULTURAL PRODUCTION TO REDUCE POVERTY, IMPROVE FOOD SECURITY AND HEALTH OF RURAL HOUSEHOLDS AND ENVIRONMENTAL CONSERVATION IN KENYA

George Ouma

Introduction
Definition of organic farming

Organic farming is a type of system that involves less intensive use of land by applying cultivation practices, which exclude or significantly restrict the use of synthetic chemicals as such as pesticides and fertilizers. Products from organic farming system are therefore defined as the technology and inputs used and not explicitly by the inherent properties of the product itself. Consumers often perceive organic product as representing an economically friendly mode of production.

Organic agriculture focuses on environmental protection and the use of natural resources, including the maintenance of soil structure and fertility, water resources biodiversity, etc. Organic farming is practised with the declared objectives of contributing to food quality and safety standards thereby protecting consumer health and facilitating international trade. It is based on technologies that prevail in parts of the developing world, which can provide employment and income for poor farmers who manage to access these market opportunities. Organic agriculture must be certified and it must adhere to globally acceptable principles which are implemented within local social, economic, climatic and cultural conditions (Mununa and LekeI, 2005). Organic farming management relies on developing biological diversity in the field to disrupt habitat for pest organism and the international maintenance of and replenishment of soil fertility (Mutsotso, 2005).

Origin and history of organic farming

In Europe, organic farming's origins in the first decades of the 20th century arose from the following events: Since the end of the 19th century the life reform movement existed and disapproved industrialization, urbanization and mechanization of the modem world. They called for a "natural way of living" including vegetarian nutrition, physical training, natural medicine and "back to the country movement". During the first and second world wars, there was a state of crisis in agriculture and agricultural science due to a drastic loss of yield despite increased use of mineral fertilizer. Farm business declined as a result of economic problems with indebtedness, compulsory activities and decline of rural tradition and lifestyle.

There were ecological problems such as breakdown of soil structure and soil fertility, decrease of seed quality, increased disease, and pest problems and increased food quality as a result of too much N fertilization (Weibel, 2002).

Organic farming for German speaking countries originated from the "natural agriculture" concept. The concept of natural agriculture included vegetarianism, healthy nutrition farming without animals and scientific (biological) understanding and soil fertility.

Cultivation included rotting and mulching techniques, conservation tillage, green manuring, rock powder fertilization, assessment of organic matter, cycling concepts for municipal wastes and human faeces (Koneman, 1939). Other concepts which promoted the development of organic farming in German speaking countries were biodynamic agriculture in which the farm was regarded as a living organism and individual characterized by "ego force" suggesting an intimate personal relationship to nature farming, organic biological agriculture concept which was characterized by a ley farming (mixed livestock and pasture), sheet composting and conservation tillage (it is farmer driven) and finally the biological technologies and methods of plant cultivation and animal husbandry (Weibel, 2002).

The development of organic farming in UK and USA was based on similar context of origin. Vegetarian food reform and back-to the land movement, scientific biological concept of soil fertility, declining soil fertility, neglected human economy, decreasing food quality, and a holistic view of nature of farming cultures of the far East. In Japan, an environmental and consumers' movement, not agriculture itself, initiated organic faming at the end of the 1960s. Urban consumers demanded healthy food and founded cooperatives to link producers and consumers.

Food production situation in Kenya

Food security is a human right declared in 1948 and reaffirmed in 1996 by the World Food Summit and the Food and Agriculture Organization of the unified nations. Over the last decade, Kenya has increasingly faced food deficits with about 50.6% of the population lacking access to food (ROK 2004, 2001). The main reason identified for food insecurity is decreased food production resulting from limited availability of productive land and increase in the frequency of drought. On the 14th July 2004 His Excellency the President of Kenya Mwai Kibaki declared the current drought a national disaster and appealed for relief food assistance. It is estimated that the government spends about US $ 40-65 million annually on famine relief. In March 2004 the government launched the strategy for revitalization of Agriculture (SRA 2004-2014) paper. It is hoped that this strategy will provide a solution to the issue of food insecurity in Kenya, which has lingered for a long time (Mutsotsi, 2005).

Land has been identified as the most important resource in agricultural production. In Kenya, limited availability of productive land is major constraint to increased agricultural thus food production. Whereas food crops are dominated by maize, which is the main staple food with an annual consumption of about 34 million bags, maize production varies between 16.6 and 34.8 million bags depending on prevailing weather conditions and producer prices thus causing frequent shortages. Over 77% of Kenyan's live in the rural areas and 18% of the land is regarded as high potential arable land. This land is highly populated resulting in an increased pressure on it. This has led to conventional agriculture to heavily rely on chemical inputs like inorganic fertilizers and pesticides raising concern on the role of these chemicals in the environmental pollution and chemical residues on the agricultural produce. This necessitates organic farming application.

The hunger task force of the UN millennium project is concerned with raising food productivity. This means the ability of small farm holders to produce more food per hectare. Sachs (2004) identified lack of adequate soil nutrients as one of the hindrances to raising food productivity. It is argued that the depletion of soils over decades and lack of access to fertilizers or biological alternatives exacerbates the problem of land scarcity. In order to increase productivity for food increase farmers must invest in soil health (Sanchez, 2004).

Extent of organic farming in Kenya

Consumer demand for organically produced food and fibre including new markets opportunities for farmers and business in developing and developed countries is increasing. Food safety concerns have pushed up the demand for organic food. World organic sales are greater than 10 billion dollars with main markets being European Union, USA and Japan. (Appendix I)

Benefits of organic farming

Improved yields

The capacity of organic resources to provide nutrients, especially N, P and K to the soils, facilitates increase in cation exchange capacity, improved water holding capacity and infiltration rates and reduced bulk density (Lakers, 2003).

Food safety and nutrition

In 2002 report by B. P. Baker, C. M Benbrook E. E. Groth III and K. C. Kenbrook indicates that organic food is far less likely to contain pesticide residue than conventional food. Pesticide residue in food may affect the nutrition value of such foods. Organic food is as safe to consume as any other food. Organic standards set strict guidelines on manure use in organic farming. Either it must be first composted or it must be applied at least 90 days before harvest which allows ample time for microbial breakdown of any pathogen.

Affordability and accessibility

Whereas subsistence farmers may face the challenge of lack of money to acquire inorganic resources, each household has access to some organic resource, most of which can be processed domestically. However, lack of information in preparation of these organic resources, a hindrance to their use and indigenous knowledge in organic resources, limits hence more research and outreach needed.

Nature of organic farming and organic resources

Organic farming builds healthy soils by nourishing the living component of the soil, and through the microbial inhabitants that release, transform and transfer nutrients form microbial inhabitants. Soil organic matters contribute to good soil structure and water holding capacity. Organic farming feeds soils biota and builds soil structure and water holding capacity with cover crops, compost and biologically based soil amendments. These produce healthy plants that are better able to resist pest and diseases organic farmers' strategy in controlling pest and disease should be prevention through plant nutrition and management. Cover crops are used and supplemented by crop rotation to change the field ecology effectively disrupting habitat for weeds, insects and disease organism. Weeds are controlled through crop rotation, mechanical tillage, and hand weeding as well as through cover crops, mulching, and flame weeding.

Organic farmers rely on diverse populations of soil organisms, beneficial insects and birds to keep pest in check. When pest population get bad, growers could implement variety of strategies such as use of insect predators, mating disruption, trap and barriers and medicinal plants. Organic farming has taken lead in agricultural practices in many countries of the world today. Approximately 2% of the US food supply is grown using organic methods. Over the last decade sales of organic products have shown an annual increase of about 20%.

In Kenya organic farming is neglected in favour of conventional farming methods. The under utilized organic resources are crop residue, agro industrial by products, domestic wastes and indigenous plants but conventional methods are not affordable to the rural poor.

Since Kenya is a poor country, the production of all necessary inputs for the maintenance of soil productivity, recycling of vegetable matters and animal extra may help to maintain soil production.

Organic farming and pesticides risk reduction

Organic farming dues not support the act of synthetic pesticides and fertilizer in agriculture and growth hormones. These chemicals have caused a lot of environmental and social problems. It has been documented that even in a highly educated country with strict training programs for farmers and other end users, poisoning costs can be high. The World Health Organization estimates that every year 3 million people suffer acute pesticides poisoning. Accidental deaths from pesticide poisoning worldwide is estimated by W.T.T.O. is 300,000 cases per year mainly in developing countries.

It has been reported that agriculture sector contributes to 14% of all occupational injuries and 10% of all fatal injuries arising from pesticides. It is estimated that almost 60-70% of all cases of acute international pesticides poisoning are due to occupational exposure worldwide. Problems of pesticide contaminated drinking water are increasing all over the world in particular developing countries. The use of pesticides is also a threat to both plants and beneficial organisms e.g. butterfly, honey, bees, earth worms, etc. Organic farming preserves and rebuilds the solid and also it preserves the endangered species. It reduces the impact of farming pollution on soil, water, etc.

The most direct benefits by the emission of pesticides are the safer working conditions for farms and farm workers. Through organic farming there will be lesser food and drinking waste contaminated with pesticides. Consuming organic produce reduce risk of dietary pesticide exposure and maximizes health benefits. Choosing organically grown food reduced the frequency and level of dietary exposure to pesticides and reduces the magnitude of risk factor that can contribute to variety of health effects (Benbrook, 2004).

Conventionally grown fresh produce is 3-4 times more likely to contain multiple residues than the corresponding organic produce. In comparison to conventional agriculture methods organic farming is considered to have less detrimental health and environmental effects.

Effects of organic farming

Organic cropping and farming systems have been researched in long-term experiments mainly in the temperate countries. A study comparing biodynamic, organic-biological conventional farming with and without farm yard manure (FYM) revealed lower yields in most organically grown crops but higher energy use efficiency and better physical, biological and chemical soil properties

(Mader *et al., 2002)*. Other studies show usually lower nitrate contents and pesticides contaminations in organically grown vegetables. Organic horticultural production is usually more labour intensive than its conventional counterparts, mainly due to mechanical weed control. This makes organic farming in horticultural production particularly suitable for overtake scale farnitly farms.

In all farming systems organic agricultural production is associated with nutrient export in form of products exported from the farm. These exports have to be balanced by nutrient inputs. Plant nutrients are bought from the decomposition of rock minerals and symbiotic fixation. Nutrition management in the organic farming system must aim at preserving nutrients from losses and making them available to the plants. Manure and compost management therefore play a central role in the organic farm.

For pest and disease management the selection of resistant varieties is important crop diversity with the field (intercropping) and between fields reduces the spread of pests and diseases. Approximate crop rotations help to reduce the multiplication of pests and diseases which are not moving fast form field to field, appropriate soil tillage buries infected harvest produce, weeds and weed seeds, facilitates infiltration of rain and creates the environment for soil organisms working the soil. (Mader *et al., 2002)*. Several workers have reported that organic farming enhances beneficial arthropods such as cerabid beetles spider numbers in the cropping system (Rosselt and Benjamin, 1994; Aster and Agis, 2000; Pfiffner, 2000; Laik *et al.*, 2000).

Certification of organic farming

In Africa, there are 2 kinds of organic farming practices. There are certified and non-certified organic farming. There is evidence of certified organic production in around 50 of African countries with approximately 320,000 ha of formerly certified land (Nicholson and Kalibani, 2004). Certified organic farming in Africa involves large and small farms.

During the year 2002 the global market for organic food and drink was valued at US $ 23 billion. Although production is increasing across the globe sales are concentrated in industrialized countries due to price premiums of the organic products. Generally buyers of organic food are those who are educated, well sensitised, with high income and those living in urban areas (Amarjit, 2004).

Certified organic farming refers to agricultural product system that has been grown and processed according to standards verified by independent accredited organizations. Establishment of guidelines for production and certification standards is an essential step in organic farming. Development of natural standards and legislation is important. Certification of organic farming ensures that the product has been produced according to defined standards and is likely to have no pesticide residues and they have grown in favourable environment. The international federation of organic agriculture movements is an organization, which provides standards that define how organic products are grown, produced, processed and handled. It provides frame work for certification bodies and standards setting organization worldwide (Lukas *et al.*, 2004).

Usually the following are allowed in organic farming:

- Fertilizers like manure, straw, green plant material, and similar materials.
- Algae, sea weed, azola
- Domestic waste and waste from parks, gardens, slaughter house, food industry, textile industry, and forest industry.
- Animal urine
- Mineral fertilizers that are in their natural conditions which are not processed. Others are row phosphate, apatite, limestone wood ash, gypsum.
- Plant protection inputs like pesticides extracted from herbs, animals, micro-organisms and insects.
- Pheromones and pheromone trap or other catching devices.
- Hot water and steam
- Pure paraffin oil and pure sulphur
- Soap ethanol and sodium bicarbonate.

Application of organic farming for subsistence horticultural production in Kenya

There are two major ways organic farming can be applied for horticultural production in Kenya.

In home gardening

Land has been identified as the most important resource in agricultural production in Kenya and limited availability of productive land is a major constraint to increase agricultura food production. Food crops are dominated by maize, which is the main staple food with an annual consumption of about 34 million bags, maize production varies between 16. 6 and 34.8 million bags depending on prevailing weather conditions and producer prices thus causing frequent shortages. There is therefore need to consider home gardening as supplementary method of food production for households. Organic resources can be best applied for home gardening.

Home gardening is not a new concept. For many generations small plots of land near the homesteads have been used as home garden (Sigot, 2001). Many communities have also practiced livestock. With time, changes have occurred in home gardening mainly as a result of land becoming scarce and the introduction of new farming technologies. These changes have largely affected the rural and pre-urban communities especially with regard to farming practices. Home gardening entails production of horticultural food crops and live stock on small pieces of land as supplementary source of food for households. Yet in many agricultural communities, people rely on one main staple food whose seasonality implies a period of food shortage causing hunger. Households can grow a mixture of foods to avoid this to prevent food scarcity during periods of drought. It would therefore be important that households diversify their food crops in home gardens in order to avoid seasonal hunger.

Raising food productivity is a concern of the Hunger Task Force of the UN Millennium Project. This means the ability of small farm holders to produce more food per hectare. Sachs (2004) identified lack of adequate soil nutrients as one of the hindrance to raising food productivity. Depletion of soil over decades and lack of access to fertilizers or biological alternatives worsens the problem of land scarcity in order to increase productivity for food insecure farmers; there must be investment in soil health in addition to other priorities (Sanchez 2004). Studies have shown that home gardens can be a sustainable means of ensuring household food security but they should be allowed to use organic farming methods. The use of green manure, compost and natural pesticides reduces food production costs and provides safe and healthy food for household consumption and even sale for income generation. Local households with well-developed home gardens possess the idea, skills and resources to produce a variety of crops and animals and to take advantage of their interdependency. The animals consume waste plants from the gardens and return nutrients to the soil in their manure. It reduces the cost of maintaining a home garden. This shows that household gardens can be developed with extremely limited economic recourses.

Traditional home garden therefore, continue to be important source of food and nutrients for rural communities and in places where land is scarce, the home garden may be the only cultivated land. FAO (2001) has reported that home garden activities are more in the wet than in the dry season. The activity of the home garden during the dry season is affected by the availability of reliable water source. Households owning land near streams and rivers utilize these waters and continue to produce horticultural crops mainly vegetable.

Organic resources would help in soil moisture conservation (Leakey, 2003). Under drought conditions field applied with manure retain moisture for a larger period. Soil moisture can also be concerned by use of organic materials such as surface mulching. Mulching also concerned moisture as well as improving soil physical condition.

To ensure food scarcity, a house garden therefore should contain a mixture of perennial and annual crops well adapted to the ecological conditions of the area. The Hunger Task Force of the UN Millennium project recognizes the important role played by a mixture of crops in improving diets. Crop mixture found in a home garden should be a good selection of a wide variety of plants and tree crops that occupy different layers play supportive roles. The crops should be intercropped. Muethoff (2001) has suggested that gardens should be highly diversified not only to include crops, domestic animals and poultry but also medicinal plants.

Fruit farming in semi-arid/Arid lands (ASALS)

More than 80% of the arable land in Kenya falls in the semi-arid and arid land where dependence on rain fed agriculture and low input application lends itself to low productivity and frequent crop failures. The arid and semi-arid land areas have excellent potential for horticultural production. Irrigation is therefore crucial in the development of the horticultural industry in these areas and can make appropriate contribution to food security and poverty alleviation (Wabule, 1999).

Farmers in the semi-arid areas of Kenya, where average annual rainfall is in the range of 500-850 mm/year, often grow maize - livestock based faming system, because maize is the main staple food in Kenya. Farmers usually intercrop beans, green grams, cowpeas sorghum and traditional crops or exotic vegetables as annual crops. Tree crop such as mangoes, paw paws, citrus, cashew nuts, macadamia are scattered on the farmland or are planted along on boundaries in the arid and semi-arid lands.

The production of annual crops such as maize is a risky enterprise due to the low and often erratic rainfall and to the occurrence of mid season droughts. These production risks prevent many farmers from applying mineral fertilizers and pesticides on their food crops and overall fertilizer use in the district is rather low (Eckert, 1999). Fertilizer is only applied in cash oriented, cotton and tobacco production systems. Since maize is the staple food crop and food self efficiency has the higher priority in decision making. Extension methods for fertilizer application have been developed to increase the yield and intensify the system.

However, few farmers who can afford to buy mineral fertilizers can follow them and do apply fertilizers to their maize and annual crop. Despite the production and marketing risks, it remains doubtful weather external input application in the ASALS is economically viable. Therefore, these farming systems can be transformed to organic system easily with tree crops products as an additional source for income generation and poverty alleviation (Ekert, 1999). Due to the production risk of annual crops, farmers have already diversified traditionally and plant many different fruit and tree crops such as mango *(Mangifera indica)*, guava *(Bidina guajava)*, Pawpaw *(carica papaya)* (Erket, 1999). The trees are nicely scattered on the farms and around homestead and are often planted along the plots boundaries as contour lines. Pruning, watering, etc., are not done. The fruits are eaten at home but some are sold in the local markets. The ASAL areas are Kitui, Machakos, Tana River, Kilifi, and Lamu Districts. The agro forestry components in these areas contribute to soil and water erosion control.

Research has discovered the potential of trees in the ASAL areas (Njenga *et al.*, 1999). Hence in these areas, more efforts should be put on the development of fruit tree crop- livestock based farming systems and to enable farmers get access to the emerging international markets. Products from organic farming can get access to markets in Canada, USA, Germany, UK and the EU.

Constraints of organic farming:

- Farmers who need to export their products must use organic certification agencies to inspect their farm annually and confirm that they adhere to standards established by various trading parties. The cost involved is high since many developing countries have no certification organizations.
- On starting organic farming a grace period of two or more years is needed to clear out chemicals on land before the crop is accepted as a product. This will delay farmers from selling of their products.
- There are very different national standards worldwide. This creates difficulty in achieving fair trade in organic production. Generally there are standards for exporting and for importing countries. The different standards may be barriers to trade.
- Difficult in getting reliable markets for organic produce abroad.
- Adverse climates in many developing countries can render storage and transport of perishable organic foods costly and difficult especially for small scale farmers who lack adequate facilities.
- Some of the products which are approved for organic agriculture like bio-pesticides and bio-fertilizers are not easily available in developing countries
- Little research has been done on natural pest control products that are recommended for organic farming.

Research needs

The research areas in organic horticultural production should be:

- Biological control of pests and beneficial behaviour
- Weed control
- Soil fertility
- Development of bio-rational pest control tools such as Kaolin, soaps, oils, repellents

Conclusion

Organic farming should be adopted in Kenya as a way of increasing food security, income, improving health and conserving the environment through growing of fruits and vegetables.

The area under organic farming in the world is as follows:

COUNTRY	AREA UNDER ORGANIC FARMING IN HECTARE (HA)
Japan	24 million ha
Australia	10 million ha
Latin America	5.8 million ha
Europe	5.5 million ha
Africa	30.000 ha
Italy	1.168.212 ha
USA	950.000 ha
UK	724.523 ha
Germany	696.978 ha
France	509.000 ha
Uganda	122.000 ha
Tanzania	55.867 ha
South Africa	45000 ha
Morocco	12.500 ha
Egypt	17.000 ha
Zambia	20.000 ha
Tunisia	18.255 ha
Kenya	494 ha
Malawi	325 ha
Benin	7359.197 ha
Zimbabwe	40 ha
Senegal	2500 ha

Source: Minouy (2004). Development and state of organic agriculture

Generally developing countries have a potential advantage in meeting demand for many organic foods in major markets due to the following factors:
(1) Organic products can grow better in tropical areas than temperate ones
(2) Developing countries have traditional production systems that do not encourage agrochemicals.

References
Amar Jit, S., 2004. Overview of the global market for organic food and drink.
Astier, M. and E. P. Agis, 2000. Evaluation of the sustainability of peasant agricultural systems, In: Thomus Alfoidi, IFAOM Scientific Conference, Basel 28 – 31, August 2000.
Eckert, M. V., 1999. In: Agong et al. (1999). Proceedings of 2nd International Horticulture Seminar on the way forward for horticulture towards Kenya's industrialization. Pp. 142 -149.
F AO, 2001. Improving nutrition through home gardening: A training package for preparing field workers in Africa, F AO.
Konemann, 1939. Biologische Bodenkultur und Düngerwirtschaft , Siebensicher Tutzing. 413 pp.
Lekas, J. K., 2003. Organic resources management in small holder agriculture. In: Organic resource management in Kenya. FORMAT, Nairobi.
Luik, A. Eenpuu, Nommiste, R., Mand K. and M. Mikk, 2000. Impact of field margins in the occurrence of beneficial insects. In: Thomus A. Foidi, William Loekeretz and Urs Niggil (eds). IFOAM 2000 - The world grows organic proceedings of the 13th International FOAM scientific conference convention centre, Basel, 28-31 August 2000. 460pp.
Mader, P. A., Fleshbach, D. and L. Dubois, 2002. Soil fertility and biodiversity in organic farming. Science, 296, 1694 -1697.
Mikou, Y., 2004. Development and state of organic agriculture worldwide.
Muelhoff, 2001. How home gardening can significantly contribute to good nutrition year round. FAO.
Mutsotsi, A. A., 2005. Utilization of organic resources to promote home gardening for food security. In: Wesonga et al. (2003). Proceedings of the fourth workshop on sustainable horticultural production in the tropies. Department of Horticulture, Moi University, Eldoret, 24th – 21th Nov 2004.
Munana, F. T. and E. Luka, 2005. Reduction of pesticides, health and environmental risks in horticultural crops through organic farming developing countries. In: Wesoga et al. (2005). Proceedings of the 4th workshop sustainable horticultural production in the tropics. Moi University, Eldoret, Kenya. 24th -21th November 2007.

108

Nicholson and F. Kilibuani, 2004. Organic agriculture in the African continent.
Njenga, A. and M. V. Ekert, 1999. The economic value of tree crops in
smallholder farming system in Kenya. In: Aganget et al., Proceedings of
the 2nd National Horticultural seminar as the way forward for horticulture
towards Kenyan industrialization. Pp. 122 -140.

PF. Filfther, L., 2000. Significance of organic farming for invertebrate diversity-
Enhancing beneficial organisms with field margins in combinations with
relationship between nature conservation biodiversity and organic
agriculture. International Workshop, Vignoliac, Italy, 1999. FOAM., Pp
52-66.

Republic of Kenya, ROK, 2004. Economic survey, 2004. Government Printer,
Nairobi.

Republic of Kenya, ROK, 2004. Strategy for revitalizing agriculture 2004-2014.
Government Printer, Nairobi.

Republic of Kenya, 2001. Report of the sector-working group on agriculture and
rural development. A poverty reduction strategy paper. Government
Printer, Nairobi.

Sachs J. D., 2004. Economic development and nutrition; How do they interact.
In: SCN news nutrition and the Millennium Development Goals.
Lavenham Press, UK.

Sanchez, P., 2004. Millennium Project Hunger Task Force. In: SCN news
nutrition and the Millennium Development Goals. Lavenham Press, UK.

Sigot, A. J., 2001. Indigenous food system. Creating and promoting sustainable
livelihoods. Paper presented at International Conference on indigenous
knowledge system: Africa Perspective, 12 - 15th September 2001.

Wabule, M., 1999. Kenya's industrialization goals through appropriate research
in horticulture. Pp. 9-13. In: Agong et al., proceedings of the 2nd National
Horticulture Seminar on the way forward for horticulture towards Kenya's
industrialization. Pp 127-140.

Weibel, F. B., 2000. Organic fruit production in Europe. Impact fruit tree.
Volume 35, No. 3

DIETARY ASCORBIC ACID SUPPLEMENTATION IN PRACTICAL DIETS FOR AFRICAN CATFISH
Clarias gariepillus (Burchell 1822) fingerlings

O. K. Gbadamosi, E. A. Fasakin and O. T. Adebayo

Abstract
The effect of dietary Ascorbic acid (Vitamin C) at inclusions levels of 0, 50, 100, 150 and 200mg AA per kg of feed on the growth performance and nutrient utilization of African catfish *Clarias gariepinus* (Burchell 1822) fingerlings were investigated in a study that lasted ten weeks. *Clarias gariepinus* weighing 6.02 ± 1.3 were randomly distributed into five treatment groups. A treatment contained three replicates of 10 fishes each. Results showed that L-ascorbic acid inclusion in the diet improved weight gain in Treatment 2, 3, 4 and 5. The highest mean weight gained (MWG) and protein efficiency ratio was recorded in treatment 4 fed a diet of 150mg AA/kg. Growth performance and nutrient utilization parameters such as specific growth rate (SGR), Feed Conversion Ratio (FCR), Protein Efficiency Ratio (PER) and Feed Efficiency (FE) in all the treatments were significantly different at $P<0.05$. After a week, 8 fish fed scorbutic diet in Treatment 1 (0mg/kg) began to develop clinical deficiency including lordosis, broken skull, pigmentation and feed refusal. In all parameters considered Treatment 4 fed 150mg AA/kg gave the best growth performance and nutrient utilization efficiencies, while fish in Treatment 1 fed scorbutic diet recorded the highest mortality of 30% and lowest growth performance and nutrient utilization efficiencies. However, fish in Treatment 5 fed the highest level of ascorbic acid supplementation of -200mg *AA/kg* did not perform better than fish in Treatment 4 fed 150mg *AA/kg* in terms of growth and nutrient utilization efficiencies.
Keywords: *Ascorbic acid, Clarias gariepinus, scorbutic, mortality, supplementation.*

Introduction
The nutritional quality of the feed is a major factor in sustaining healthy fish. It has been shown that the immune system of fish can be enhanced by the used of immuno modulators and antioxidants vitamins like Ascorbic acid (AA) (Verlhac and Gabandan, 1997). The combination of good management and nutritional prophylaxis will produce better survival rates and improve growth in aquaculture (Fasakin, 1997).

The inability of many fish species to synthesize ascorbic acid (AA) or Vitamin C, which is essential for fish growth, health and reproduction, is well documented (Dabrowski, 1990; Soliman et al., 1986). Ascorbic acid must therefore be supplied via feed. The symptoms associated with ascorbic acid deficiency are also well documented for many cultured fish (Halver, 1989; Sadnes et al., 1992).

Due to the multiple role of ascorbic acid in various metabolic pathways, a better understanding of the mechanism through which ascorbic acid as a nutritional element influences the immunological systems in modem intensive fish farming is necessary. It is therefore prudent to investigate the growth performance, nutrient utilization and to establish the ascorbic acid requirement on species by species basis.

Many authors have established the ascorbic requirement of some species of commercial importance. This has been summarized by Halver et al., (1990). However, because of disparities in methodology and assessment criteria, there are considerable differences both between and within species in term of proposed requirement. Natural ascorbic acid is unstable and feed manufacturer must add more than the requirement to ensure adequate levels in feed processing. Recent studies indicated that inclusion of phosphate derivatives (Ascorbate -2- polyphosphate) were resistant to oxidation and retained ascorbic acid activity for fish (Abdelghany, 1996). The clariid catfish *Clarias gariepinus* (Burchell, 1822) is the most important fish species cultured in Nigeria, where the study was undertaken. This specie has shown considerable potential as a fish suitable for intensive aquaculture (Balogun et al., 1992). This fish grows rapidly, is resistant to disease and stress, sturdy and highly productive even in poly-culture with many other fish species like Nile Tilapia *Oreochromis niloticus* (Fagbenro et al., 1997). However, there is dearth of work in both qualitative and quantitative ascorbic acid requirement of the African catfish *Clarias gariepinus*.

This study is therefore aimed at evaluating the optimum growth performance and nutrient utilization efficiencies of African catfish, *Clarias gariepinus* fed dietary L-ascorbic acid supplemented diet.

Materials and methods
Experimental diets

Five isocalorific and isonitrogenous diets containing 40% crude protein and 12% lipid were formulated for fingerlings catfish, *Clarias gariepinus* in a ten-week trial experiment (Table 1). Ascorbic acid, commercially available as ROVIMIX STAY C- (Roche, Istanbul, Turkey) was used. Diet without ascorbic acid supplementation served as the control. Ascorbic acid supplementation in

111

diets 2 to 5 were 50.0, 100.0, 150.0, 200.0mg/kg respectively. All dietary ingredients were first milled to small particle size (approximately 250mm), ingredients including ascorbic acid were thoroughly mixed in a Hobart A-200 pelleting and mixing machine (Hobart Manufacturing Ltd., London, England) to obtain a homogenous mixture, cassava starch was used as a binder. The resistant mash was then passed without steam through a 0.9mm die to obtain five stands which were sun dried immediately. Diets were broken up and sieved to convenient sizes and stored at (-18°C) prior to feeding.

Experimental fish and management
C. *gariepinus* fingerlings with average weight of 6.0±0.4g were randomly distributed into glass tanks (60cm x 45cm x 45cm) at ten fish per tank. Each treatment was in triplicates group of fish. Tanks supplied water from a borehole powered by 1.5HP pumping machine. Water temperature was maintained at 24±0.5 dissolved oxygen was kept at a saturation level of 6±0.1. The fish were fed acclimated to the experimental conditions and their respective diets for two weeks prior to the start of the feeding trial. The fish were fed with their respective diets at 5% body weight twice daily at 9.00 and 16.00 hours throughout the duration of the experiment. Fish weights were determined at the 7th day of each week and the quantity of feed adjusted based on the changes in body weight of fish for subsequent feeding.

Proximate composition
Proximate composition of diets and fish carcasses before and after experiment was performed according to AOAC (1990) for moisture content, fat, fibre and ash. Ascorbic acid was determined by semi-automated flourometric method as described by AOAC (1990).

Performance Evaluation
Fish performances during the experiment were based on productivity indices on growth performance and nutrient utilization efficiencies as described by Fasakin *et al.,* (2003) as follows:

- Mean weight gain (MWG): Mean final weight - Initial weight
- Total percentage weight gain (TPWG %), total weight gain/Initial weight x 100
- Specific growth rate (SGR), loge Wt - Loge WD / T x 100
- Where Wt = Final weight (g), W_1 = Initial weight (g) and T = rearing periods (days)
- Feed conversion ratio (FCR) = (dry weight gain (g) / fish weight gain (g)

- Protein efficiency ratio (PER) = (fish weight gain (g) / protein fed (g))
- Feed efficiency = Live weight gain (g) / feed supplied (g)

Data Analysis

Biological data generated were subjected to one-way analysis of variance (ANOVA). Where means were significantly different, they were compared with Duncan's multiple range test (Zar, 1984).

Results

Results obtained in this study were as shown in Tables 1, 2 and 3. Table 2 shows the performance evaluation indices of C. *gariepinus* fed on the test diets. The results show that the productivity rates of the fish (SGR, WG, FCR, PER, FE) were significantly different (P<0.05). Comparatively, fish fed on scorbutic (control) showed drastic reduction in weight and significantly different (P < 0.05) from other groups of fish fed ascorbic acid supplemented diets.

Fish fed 150mg/kg ascorbic acid shows the highest % weight gain and specific growth rate of 212.12% ±1.16 and 1.61±1.20 respectively. The best feed conversion ratio (as fed basis) was also found in diet 4 which contained 150mg/kg ascorbic acid supplementation. However, highest mortality was recorded in Treatment one (Control) fed no ascorbic and supplementation as 30%, followed by Treatment two (50mg/kg) ascorbic acid supplementation while no mortality was recorded in Treatment 3, 4 and 5 which had 100, 150 and 200mg/kg respectively.

The results of the whole body proximate composition of the fish at the beginning and the end of the experimental period are presented in Table 3. The protein and lipid contents of fish showed a marked increase over the initial whole body composition, although, the protein values of fish were similar in all the treatments, there were significant difference (P < 0.05) in protein values of fish in Treatment one fed scorbutic diet which showed the lower values of protein than fish fed on ascorbic acid supplemented diet.

Discussion

The result of this study shows the efficacy of ascorbic acid supplementation in the diets for the African clariid catfish C. *gariepinus*. Fish fed scorbutic diets showed dietary related mortality and morphological symptoms such as impaired collagen formations. The temperature (24.6 ± 0.1) and dissolved oxygen (6.68 ± 0.1mg/L) values are within the range recommended for African Catfish culture Boyd (1986).

Fish fed 150mg/kg AA supplemented diets showed the best growth profile during the feeding period. Fish in treatment one fed scorbutic diet showed reduction in growth rate and weight gain from the 8th week, this agreed with the results recorded by Li and Lovell(1985) for Channel catfish fed scorbutic diets. This could be associated to the ascorbic acid variation in the diets, which was the only heterogeneous factor in the experimental diets. However, fish in treatment 5 fed 200mg/kg Ascorbic acid did not perform better than treatment 4, which was fed 150mg/kg Ascorbic acid supplemented diets this agreed with the Halver *et al.*, (1989) that there was no effect of feeding a megadose of Ascorbic acid supplemented diet in Channel catfish than required by the fish. The mechanism behind this effect is postulated to be decrease synthesis of bile from cholesterol in Ascorbic acid deficiency through inhibition of the rate of limiting enzyme cholesterol-7-a-hydroxylase (Li *et al.*, .1993).

Nutrient utilization indexes (BWG, SGR, FCR, and PER) in fish fed scorbutic diets was poor as was also recorded by Baker *et al.*, (1998) in diets of Channel catfish fed ascorbic acid supplemented diets, the nutrient utilization of fish in treatment 2 to 4 increased with increasing level of ascorbic acid supplementation resulting in an improved nutrient utilisations of the catfish. The trend of performance agrees favourably with the findings of Sadnes *et al.* (1992), Baker and Davies (1998), Odubanjo (2001), that ascorbic acid, a-tocopherol, and niacin supplementation improves growth performance and nutrient utilization of Clarias gariepinus.

Conclusion

A dose of 150mg/kg ascorbic acid supplementation is therefore recommended in the diet of African Catfish, Clarias gariepinus, this is higher than the NRC (1993) recommendation of 100mg/kg for Channel catfish. However, the role of ascorbic acid in free radicals protection and its interaction with other antioxidants like vitamin E should be studied in fish nutrition.

Appendix

Table 1: The experimental diet composition in g/100g dry matter containing various inclusion levels of ascorbic acid supplementation for *Clarias gariepinus*

Ingredients	Treatments				
	1 control	2	3	4	5
Fish meal (70%)	22.00	22.00	22.00	22.00	22.00
GNC	28.00	28.00	28.00	28.00	28.00
SBM	24.00	24.00	24.00	24.00	24.00
Yellow maize	11.00	11.00	11.00	11.00	11.00
Vegetable oil	5.00	5.00	5.00	5.00	5.00
Oyster shell	2.00	2.00	2.00	2.00	2.00
Rice bran	4.00	4.00	4.00	4.00	4.00
*Vit/Min premix	2.00	2.00	2.00	2.00	2.00
Salt	1.00	1.00	1.00	1.00	1.00
Starch	1.00	1.00	1.00	1.00	1.00
Ascorbic acid mg/kg		50	100	150	200

Premix as supplied by Animal Care, Limited, Lagos, Nigeria.
Vitamins supplied mg/l00g diet: thiamine (B_1) 2.5mg: riboflavin (B_2), 2.5mg pyridoxine 2.0mg: pantothenic acid, 5.0mg: inositol, 3mg: biotin, 0.3mg: folic acid, 0.75mg para-amino benzoic, 2.5mg: chlorine, 200mg; niacin, 10.0mg, cycobalamin (B_{12}), 10.0mg; menadione (k), 2.0mg. Minerals: $CaHPO_4$, 727.8mg: Mg SO_4, 1275mg , 60.0mg; kcl 50.0mg; $FeSO_4$, 250mg, $ZnSO_4$, 5.5mg; Mn_4SO_4, 2.5mg $CUSO_4$, 0.79mg; $CoSO_4$, 0.48mg; $CaClO_3$, 0.3mg; Cr Cl_3.

Table 2: Proximate composition of experimental diets (%DM)

	T_1	T_2	T_3	T_4	T_5
Crude protein	40.28	40.19	40.21	40.13	40.09
Lipid	12.39	12.21	12.33	12.17	12.03
Crude fibre	5.09	5.28	5.11	5.19	5.42
Ash	8.35	8.36	8.48	8.33	8.44
Moisture content	13.41	13.54	13.61	13.48	13.37
Nitrogen-free extract (NFE)[1]	20.48	20.42	20.26	20.70	20.65
Added ascorbic acid mg/kg	0.0	50.00	100.00	150.00	200.00
Measured ascorbic acid mg/kg	0.64	56.20	109.70	165.90	204.83
Gross energy[2] kcal/100g	431.30	429.00	429.40	429.70	427.50

[1]Nitrogen-free extract: calculated as 100- (crude protein + ash + crude fibre + ether)

[2]Gross energy (kcal/100g) based on 5.7kcal protein; 9.5kcal/g lipid; 4.1kcal/g carbohydrate.

Table 4: Proximate composition (% wet weight) of the carcass of *Clarias gariepinus* fed experimental diets containing varying inclusion levels of ascorbic acid

Parameters	Sample Initial %	Final sample of Fish					S.E.
		T_3	T_1	T_2	T_4	T_5	
Moisture	73.05	71.50[a]	73.01[c]	72.18[b]	71.39[a]	72.11[b]	0.16
Protein	14.15[a]	17.14[d]	15.23[b]	16.15[c]	17.16[d]	17.11[d]	0.28
Fat	6.18[b]	7.21[a]	7.03[a]	6.40[a]	7.19[a]	6.91[b]	0.08
Ash	4.70[a]	4.24[bc]	4.33[b]	4.56[ab]	4.12[cd]	4.07[d]	0.93

Figures in each row having the same superscripts are not significantly different ($P > 0.05$)

Table 3: Cumulative growth performance, nutrient utilization and cost index of *Clarias gariepinus* fed varying levels of ascorbic acid

PARAMETERS	Treatment 0	Treatment 1	Treatment 2	Treatment 3	Treatment 4
Final weight	8.81±0.10[a]	10.12±1.40[b]	12.61±1.10[c]	17.01±0.10[e]	14.76±1.20[d]
Initial weight (g)	5.94±0.30[a]	6.02±0.50[a]	6.06±0.40[a]	6.05±0.30[a]	6.02±0.20[a]
Weight gain (g)	2.87±0.50[a]	4.10±0.50[b]	6.55±0.15[c]	10.96±0.17[e]	5.74±0.22[d]
Daily weight gain (g)	0.04±0.02[a]	0.06±0.02[b]	0.10±0.02[c]	0.16±0.60[e]	0.12±0.03[d]
Feed fed (g)	12.89±0.18[a]	17.50±0.24[b]	19.39±0.21[c]	21.70±0.41[e]	20.13±0.16[d322]
Daily feed int. (g)	0.18±0.18[a]	0.21±0.24[b]	0.2S±0.19[c]	0.31±0.38[e]·	0.29±0.15[d]
FCR	4049±0.02 a	4.27±0.01[b]	2.96±0.03[c]	1.98±0.08[e]·	2.30±0.04[d]
SGR	0.S6±0.06 a	0.74±0.04[b]	1.05±0.05 c	1.48±0.02[e]	1.28±0.15[d]
% Weight gain	48.32±0.71[a]	68.11±0.65[b]	108.09±0.9[c]	181.16± 1.5[e]	145.18±0.89[f]
Feed efficiency	0.22±0.12[a]	0.23±0.13[a]	0.39±0.10[b]	0.51±0.14[c]	0.43±0.11[c]
Hepatosomatic index	0.05±0.13[a]	0.08±0.05[a]	0.26±1.03[b]	1.39±0.05[d]	1.01±0.04[c]
Protein intake (g)	5.19±0.16 a	7.03±0.21[b]	7.80±0.12[bc]	8.71±0.12[c]	8.07±0.10[c]
PER	0.55±0.01a	0.58±0.01[b]	0.84±0.02[c]	1.26±0.01[e]	1.08±0.02[d]
Pellet stability %	96.01 a	95.52 a	94.55 a	97.10 a	93.82[a]
Cost index N/kg	170.50 a	172.50 a	175.0 a	177.50 a	180[a]
Mortality %	30 a	10[b]	c –	c –	c –

Figures in each row having the same superscripts are not significantly different (P>5)

References

Abdelghany, A. E., 1996. Growth response of Nile tilapia *Oreochromis niloticus* to dietary L-ascorbic acid, L-ascorbnyl-2-sulfate, and L-ascorbyl-2-polyphosphate. J. World Aquaculture. Soc., 1996. 27:449-455.

AOAC, 1990. Association of Official Analytical Chemists. Official Method of Analysis. (15 ed), vol. 1. K. Heltich (ed.), Arlington, Virginia.

Baker, R. T. M., Morris, P. C. and S. J. Davies, 1998. Nicotinic acid supplementation of diets of the African catfish, *Clarias gariepinus* (Burchell). J. Aq. Research, pp 29: 791-799.

Balogun, A. M. and K. Dabrowski, 1992. Improvement in the nutritive quality of soybean meal by co-ensiling with under-utilized fish discards. Proc. Fish Soc. Nig., 10:75-86.

Boyd, C. E., 1986. Water quality in warm water fish ponds. Agricultural Experiment Journal. Aubum University 35pp.

Dabrowski, K., 1990. Ascorbic acid status in the early life of white fish *(Coreyonus lavaretus L.)*. Aquaculture 84, 61-70.

Fagbenro, O. A., Balogun, A. M. and Fasakin, E. A., 1997. Dietary methionine requirement of the African catfish *Clarias gariepinus*. J. of Applied Aquaculture (In Press).

Fasakin, E. A., 1997. Studies on the use of Azolla africana DESV (WATER FERN) and spirodela polyrrhiza L. Schleiden (Duckweed) as feedstuffs in production diets for *Orechromis niloticus* and *Clarias gariepinus* fingerlings. Ph.D thesis, Dept. Fisheries and Wildlife, Federal University of Technology, Akure, Ondo State, Nigeria.

Halver, J. E., 1989. The Vitamins. In: Fish Nutrition, 2nd ed (J. E. Halver, ed.), pp 32-102. Academic Press New York. NY.

Li, Y. and R. T. Lovell, 1993. Vitamin C and disease resistance in channel catfish *(Ictalurus punctatus)*, Can. J. Fish Aquat. Sci. 39:948-951.

NRC, 1993. (National Research Council). Nutrient Requirement Fish. National Academy Press. Washington, D. C, 114pp.

Sadnes, K., Torrisen, O. and R. Waagbo. 1992. The minimum dietary requirement of Vitamin C in Atlantic salmon (Salmon salar) fry using Ca ascorbate -2-monophosphate as dietary source. Fish phiscol. Biochem. 10:315-319.

Soliman, A. K., Jauncey, K. and R. J. Roberts, 1986. The effect of varying forms of dietary ascorbic acid on the nutrition of juvenile tilapias *(Oreochromis niloticus)*. Aquaculture: 52: 1-10.

Odubanjo, A. B. The utilization and growth response of African Catfish *(Clarias gariepinus)* fingerlings fed diets with varying levels of livestock vitamin premix (Growers) inclusions. A project submitted to Department of Wildlife and Fisheries Management, University of Ibadan, Nigeria.

Verlhac, V. and Gabaudan, 1997. The effect of Vitamin C on fish health. www.roche.com /nutrition.

Zar, J. H., 1984. Biostatistical analysis. Prentice-Hall, Englewood Cliffs, New Jersey, USA.

MANAGEMENT STRATEGIES TO FOSTER SUSTAINABILITY IN THE POULTRY INDUSTRY IN SUB-SAHARAN AFRICA

Victoria N. Meremikwu

Summary

Several factors have been identified as constraints to sustainable poultry production in sub-Saharan Africa. (1) Technology transfer of poultry production systems in environments that are unsuitable for such productions. (2) Economic crises due to foreign currency difficulties which affected the importation of feedstuffs and grand-parent stock. (3) Technical constraints which include (i) poor chick quality due to delayed access to feed post-hatch and (ii) nutritional standards e. g, NRC, (1994) which may over-specify in this region. Strategies are provided to contain the prevailing constraints while exploiting the genetic potentials of the birds using available resources. Matching poultry production with available feed resources by the use of alternative feedstuffs is provided as the key strategy. Index is provided for use in the choice of substitute feedstuffs so as to strike a balance between cost and quality of feed. Methods of including alternative feedstuffs in rations formulation without compromising bird performance are provided in the following strategies: "Ration formulation flexibility" and "Use of diets formulated to lower nutrient specifications". Early feeding of hatchlings in the hatchery immediately post-hatch is provided as a strategy to improve the functional capacity of the birds. The benefits of these strategies are summarized in the concluding remarks.

Introduction

The poultry industry in sub-Saharan African countries has seen a steady and progressive increase in the prices of production inputs and poultry products during the last twenty years. This is because, for most of the period before the inflationary session, there has been a surplus of food grains in the Western countries and the pace of importation of grains and chicks from these countries exploded into gigantic volume of international trade such that no consideration was given to the issue of sustainability as it affects feeds and feeding of poultry (Obioha, 1992). According to Smith (2001) sustainability in the poultry industry can be obtained only when a country produces a large surplus of vegetable foodstuffs over and above the needs of the people, otherwise, intensive poultry will become a liability rather than an asset.

It has been observed that many countries in the developing world have no longer the luxury of importing feed ingredients particularly maize and soybean (which are generally accepted as the basis for poultry diets) because of foreign currency difficulties. Most of the countries in these regions do not have the land or climate to grow these crops.

Those of them that did import these ingredients have been in precarious situations as a result of economic crises such that the poultry industry has collapsed. For example, nearly 70% of all poultry farms and 50% of feed manufacturers had ceased operation in Indonesia (the hardest hit of all countries) as a result of foreign currency difficulties (Hutagalung, 1998). According to Udedibie (2003), the Nigerian poultry industry was rocked by crises as an indirect result of the ban on importation of maize by the Federal Government.

The economic crises go further than just the feeding of poultry and extend into the importation of grand-parent stock to produce high-performing layers and broilers. Many if not all the countries in sub-Saharan Africa cannot develop their own strain of poultry breeding stock partly because they lack the genetic resources to match the performance of imported breeds and partly because of outside pressure not to do so (Farrell, 2005). Low performing poultry lines are being offered by breeders to the market and these are being used for certified poultry production in many of the sub-Saharan African countries (Flock et al., 2005). The low performance of these strains of poultry is not compatible with sustainability objectives as it involves waste of resources (food, water and energy).

Again, researchers in sub-Saharan Africa did not have the opportunity to investigate the extent of successful adaptation and performance of every imported strain of bird under the local climatic conditions as a result of high rate of importation of exotic chicks from Western countries during the period of technology transfer. Undeniably, a major cause of most pressing constraints in project execution in the third world has been the transfer and use of technologies which are totally inappropriate to prevailing conditions. Recent studies by Gous and Morris (2005) revealed that the low performance of broilers in sub-Saharan African regions is because broiler production in the tropics is being introduced as a farming system in environments that are unsuitable for such operation. A typical example is a study by Cresswell (2004) where the performance of broiler is compared in trials conducted in several countries against standard values for the strain of birds used. These countries in the humid tropics had generally poorer all-round performance with higher feed cost compared to that in the more temperate climate. According to this author, the lower expected performance of these birds reflects the effects of the harsh tropical climate of sub-Saharan Africa.

Apart from environmental and economic constraints, there are other important constraints that go unnoticed but they are eating deep into the resources of the poultry industry in sub-Saharan Africa. These are grouped under technical constraints and include: chick size and nutritional standards.

The purpose of this paper is to highlight some of the effects of these constraints and provide documented strategies to foster sustainability in the poultry industry in this region. The aim is to incorporate recent innovations that originated from scientific research into the feeding of poultry using available resources.

The objective is to sustain poultry production in sub-Saharan Africa under the prevailing constraints by exploiting the biological and genetic potentials of the birds from relatively cheap nutrient sources.

Technical constraints

Chick size

Delayed access to feed post-hatch has profound adverse effect on the productive life of birds. According to Uni and Ferket (2004) poor chick and poultry quality, due to delayed access to feed post-hatch costs, the poultry industry loses annually $200 million as a result of efficiency and product yield. Under practical conditions, most chick are given access to feed only 36-48 hours after hatching, which results in the mobilization of body reserves to support metabolism and thermal regulation. This decreases bodyweight and impairs overall performance (Noy and Sklan, 1999). Such hatchings become more susceptible to disease pathogens, they lose weight and the development of critical tissues is restricted. According to Uni and Ferket (2004), post-hatch mortality in these hatchings is up to 5% and many survivors exhibit stunted growth, inefficient feed utilization, reduced disease resistance and poor meat yield. Noy et al. (2001) reported that the loss in body weight and impairment of overall performance due to delayed access to feed post-hatch is perpetuated through to marketing.

Nutritional standards

Nutritional standards are provided to guide feed manufacturers during ration formulation. For example, Smith (1990) emphasized that feed manufacturers should not be interested in producing the cheapest ration but one which will give the greatest return per unit cost spent on food. Even with the recognition that it is not possible in many low income, resources poor countries (particularly those in the humid tropics) to obtain bird performance observed in many Western countries because of environmental constraints, yet nutritionists in these areas insist on using nutrient requirements (e.g. NRC, 1994) which may over specify in these regions, thus leading to waste of resources. Again, utilization of conventional feedstuffs to meet nutritional standards in poultry rations places a serious burden on the overall food supplies of the countries in sub-Saharan Africa. This is due to stiff competition with channels in the food chain which include: human food channel, industrial channels, and export channels (both official and smuggled). These channels command higher priority and can pay higher prices (Esonu, 2000).

Strategies of fostering sustain ability in the poultry industry in sub-Saharan Africa

Use of alternative feedstuff

The ability to utilize alternative feedstuff in commercial poultry rations is the first and most important step towards sustainable production in sub-Saharan Africa. This is because alternative feedstuffs allow poultry producers to match their productions with available resources thereby sustaining production even in the midst of economic crises. According to Farrell (2005), the prospect of developing a stable and sustainable poultry production that depends on matching poultry production with available resources, although daunting, must be done. Hutagalung (2000) had stated succinctly that the economic crises in many countries in sub-Saharan Africa has taught them a very costly lesson for being complacent and placing too heavy a dependence on imported raw materials. This means that such countries must match their production with available feed resources.

Apparent index has been provided by Obioha (1992) in the choice of substitute feedstuff so as to strike a balance between cost and quality of feed. The index is based on expression of cost of feedstuffs in price (e.g. Naira) per percentage protein content. Protein is used as a yardstick because it is the most limiting nutrient, the most expensive nutrient and the best indicator of diet quality. If a conventional feedstuff such as soybean meal costs 10,000.00 Naira per ton and contains 50% protein, its cost per percentage protein content is 200.00 Naira. This index can then be used to compare with substitutes provided that their amino acid profile, fat and fiber content are comparable. Esonu (2000) gave a similar index. According to this author, in situations where a rather limited list of feedstuffs is feasible to use, usually a selection is made from those that provide the major nutrients at the cheapest cost per unit of energy or protein without compromising performance.

Ration formulation flexibility

Ration formulation flexibility is a concept that involves the use of a range of feed ingredients which normally would not be included in conventional poultry rations. It encourages the use of alternative feed resources. According to Obioha (1992), the success of any feed industry is the ability to achieve the irreconcilable targets of using the cheapest available ingredients to achieve the highest quality mixed feed. This trend in ration formulation had earlier been suggested by Bercovici *et al.* (1989), "when looking at least feed cost per unit of product produced, nutritionists should put their major emphasis on formulating to produce the most economical product regardless of feed consumption rather than trying to meet a rigid set of specifications".

The commercialization of this type of ration formulation is made very easy by the use of the spreadsheet (a computer program which is available in almost every computer in the developing countries). Such a program can be used to set up a database (in this case a table containing information about both the nutrient content of feedstuffs and the nutrient requirements of one or more types of poultry). Once the data have been entered and the formula added, then the different rations and their nutrient contents can be calculated automatically. The importance of the spreadsheet in these types of ration formulation is that the different rations can be recalculated by merely changing the percentage content of the diet (i.e. the proportion of individual feedstuff) and the spreadsheet will automatically recalculate the nutrient content of the new diet. One major advantage of the spreadsheet in this case is the possibility of adding the prices of the ingredients as one of the columns so that the price of the diet is calculated automatically. This information is a valuable management aid as it tells the nutritionists the price to pay for alternative feedstuff (see Appendix 1).

Use of diets formulated to lower specifications

The recognition that production conditions in many countries in sub-Saharan Africa do not allow poultry to teach their full genetic potentials irrespective of the level of input implies that comparing performance parameters (growth rate, feed efficiency and egg number) with those in Western countries cannot be affordable. According to Farrell (2005), the concept of maximizing growth rate of meat birds and laying hens must be matched with economic consideration. This calls for the use of diets formulated to lower nutrient specifications than would otherwise be possible. This concept may compromise bird performance (growth and egg production) but in many sub-Saharan Africa countries, this may be the most economical choice. This is because, when lowering nutrient specifications, more of the poorer quality feeds ruffs can be brought into ration formulation and this will allow producers to match their production with available feed resources.

Lowering available metabolize able energy (AME) value of poultry ration
The capacity of older birds to increase feed intake in response to diet dilution (i.e. reducing AME content) allows a high inclusion of low quality feeds ruffs for the later weeks of age, thus allowing the use of a wider range of ingredients which would not be used in conventional finisher formulation. Leeson *et al.* (1992) demonstrated the remarkable capacity of broilers aged 35 days old to increase feed intake in response to diet dilution without altering substantially body weight and breast meat yield. A recent experiment carried out at the National Research Institute in Papua New Guinea (PNG) showed that it was possible to dilute a broiler finisher diet with 400g coconut meal/kg without loss of performance (Table 2).

The ability to digest feedstuffs increases as poultry ages. This phenomenon had earlier been recognized by Johnson (1987) when he fed the same ingredients

Table 2. Dilution (%) of a broiler finisher diet (BF) with coconut meal (CM) on broiler performance from 21-53 days of age.
Values followed by the same letter are not different (P<0.05).

Diet (%)		Protein (%)	Intake (kg)	Weight gain (kg)	Final weight (kg)	FCR	Cost (k/kg)
BF 1000		23.3	4.20	1.665$_a$	2.191$_a$	2.53$_{ab}$	6.18$_a$
BF80+20CM		20.1	4.02	1.681$_a$	2.208$_a$	2.40$_a$	5.12$_b$
BF60+40CM		19.6	4.03	1.645$_a$	2.168$_a$	2.45$_a$	4.39$_c$
BF40+60CM		18.9	3.39	1300$_b$	1.823$_a$	2.61$_b$	3.81$_d$
BF2Q+80CM		19.3	2.48	0.702$_c$	1.22$_{ba}$	3.55$_c$	3.84$_d$
LSD			0.169	0.102	0.104	0.157	0.227

Source: Farrell (2005)

to young and old birds. He observed that the AME values were higher for the older birds. This may lead to the establishment of the concept of "priority scheme of feedstuffs" in which the high quality ingredients will have priority in broiler starter rations while the lower quality ingredients may be used for the older birds.

Use of low dietary protein sources

There are several good reasons why poultry should be fed on little or no protein concentrate in sub-Saharan Africa;

1. Nitrogen excretion will be reduced, thereby reducing the energy cost of excreting surplus nitrogen. This might help the birds to cope with heat stress (Gous and Morris, 2005).

2. Protein sources are both scarce and expensive in the developing countries.

Studies had been carried out by Farrell and Martins (1998) in which some of the poultry rations had no added protein sources but the deficits were made up of free-amino acids following analyses of ingredients and adjusting for their digestibility. The control ration was formulated with conventional ingredients. All rations gave similar production except that nitrogen excretion was lower in all rations compared to the control. According to Bervovici *el al.* (1989) the adequate supply of amino acids required by advanced animal nutrition is difficult to achieve when only vegetable and animal protein sources are available. Industrial amino acids are competitively available and can replace protein rich sources to match amino acid requirements avoiding dietary protein sources (Farrell and Martins, 1998).

In a case where only "poor quality" vegetable protein sources are available and amino acid supplements are either unobtainable or prohibitively expensive, the sensible practical precaution is to enter all specifications for essential amino acids (Eaas) as proportions of the protein content of the ration. A minimum value for crude protein concentration should then be set to avoid the risk of formulating a "well-balanced" but inadequate ration. Otherwise, with no maximum set for protein in a least-cost feed formulation, the computer will increase the use of cheap natural proteins to meet the requirement value entered for Eaa. In the case of growing birds, this can lead to an imbalance between the first limiting amino acid and the background level of other amino acids, resulting in depressed performance (Morris, 2004).

Early feeding of hatchling

The limitations by delayed access to feed post-hatch can be alleviated by the administration of food in the hatchery immediately pest-hatch, a technology termed "Early Feeding" (Uni and Ferket, 2004). Providing caloric nutrients in solid or liquid form resulted in considerable increase in body weight, which was observed to be greatest between 4-8 days and diminished thereafter (Noy and Sklan, 1999). The higher performance of the birds that had feed immediately post-hatch over their counterparts that were held without feed was as a result of the efficient development of the intestine. According to Geyra *et al.* (2001a), the extensive changes in the morphological development of the gut that occur at hatch are apparently dependent on the first access to fed post-hatch and sensitive to delay in nutrient supply. The intestine is the primary nutrient supply organ, the sooner it achieves the functional capacity, the sooner and better the young bird can utilize dietary nutrients and efficiency grow at its genetic potential and resist infections and metabolic diseases. One important effect of early feeding of hatchlings post-hatch is that the effect is perpetuated through to marketing (Noy and Sklan, 2001). At marketing, all broilers that received early access to nutrients immediately after hatch were 8-10% heavier than those that were held without feed or those that had just received water. Studies through to marketing in poultry showed trends similar to chicks. The percentage of breast muscle of broilers and turkeys at marketing were higher by 7-9% and 4-10% respectively.

Concluding Remarks

Alternative feedstuffs allow poultry producers to match their productions with available resources thereby sustaining productions even in the midst of economic crises.

Ration formulation flexibility enhances the use of alternative feedstuffs because it involves the inclusion of a wide range of ingredients which normally would not be used in conventional poultry rations. The computer spreadsheet offers opportunity for automatic calculation of nutrient contents and price of diets thereby informing the nutritionist the price to pay for substitute feedstuffs.

The concept of lowering nutrient specification may compromise bird performance, but in many sub-Saharan African countries this may be the most economical choice. This is because maximum performance is unlikely to be achieved in these countries irrespective of the level of inputs, because of environmental constraints. The concept of maximizing growth rate of meat birds and laying hens must be matched with economic consideration. This is important, because when lowering nutrient specifications, more of the poorer quality feedstuffs can be brought into ration formulation and this will allow producers to match their production with available resources.

Early feeding of hatchlings in the hatchery immediately posts hatch improves the functional capacity of the birds because of the efficient development of the intestine. This allows the birds to efficiently grow at their genetic potentials because of efficient utilization of dietary nutrients.

(Appendix 1)

See example of using the spreadsheet in Table 1.

Table 1. Calculating the dietary requirements of a laying hen using a spreadsheet

Foodstuffs	c	ME (kJ/g)	Protein (%)	Calcium (%)	Phosphorus (%)	Price $/ton
		D	**E**	**F**	**G**	**H**
8 Maize-meal		14.5	8	0	0.4	150
9 Fish-meal		13	56	5	2.5	300
10 Soya bean meal		11	44	0.25	0.6	250
11 Limestone flour		0	0	35	0	50
12 Bone-meal		0	0	22	9.0	70

Foodstuffs	% in diet	ME (kJ/g)	Protein (%)	Calcium (%)	Phosphorus (%)	Price $/ton
	C	**D**	**E**	**F**	**G**	**H**
1 maize	60	=D8*C21/100	=E8*C21/100	=F8*C21/100	=G8*C21/100	=C21*H8/100
22 fish-meal	10	=D9*C22/100	=E9*C22/100	=F9*C22/100	=G9*22/100	=C22*H9/100
23 Soya bean meal	10	=D10*C23/100	=E10*C23/100	=F10*C23/100	=G10*C23/100	=C23*H10/100
24 Limestone flour	10	=D11*C24/100	=E11*C24/100	=F11*C24/100	=G11*C24/100	=C24*H11/100
25 Bone-meal	C21:C 25	=D12*C25/100	=E12*C25/100	=F12*C25/100	=G12*C25/100	=C25*H12/100
26						H21:H25
27 Total nutrients		9D21:D21:=D27	E21:E25	F21:F25	G21:G25	
SUM 28	100					
29 Estimated Requirements Of a laying hen		12	17	3	0,8	
Deficit or surplus nutrients		D27-D29	E27 –E29	F27-F29	G27-G29	

Source: Smith (2001)

References

Bercovici, D., Gaertner, H. F. and A. P. Puigserver, 1989. Poly - L- Lysine and L- Methionine as nutritional sources of essential amino acids, Journal of Agric. and Food Chemistry 37:873-877.

Cresswell, D., 2004. Broiler performance and production part 3. Economic implications. Asian Poultry. August: 26-29.

Esonu, B. O., 2000. Animal Nutrition and Feeding - A functional approach, Rukzeal and Ruksons Associates Memory Press, 21 Rotibi street, Owerri, Imo State, Nigeria. Pp 1-224.

Farrell, D. J., 2005. Matching poultry production with available feed resources. World's Poultry Science Journal 61: 298-307. No. 2.

Farrell, D. J and E. A. Martin. 1998. Strategies to improve the nutritive value of rice bran in poultry diets. British Poultry Science 39:601-611.

Flock, D. K., Laughlin, K. F. and J. Bentley, 2005. Minimizing losses in poultry breeding and production, how breeding companies contribute to poultry welfare. World's Poultry Science Journal 61:227-237 No. 2.

Geyra, A., Uni, Z. and D. Sklan, 2001a. The effect of fasting at different ageson growth and tissue dynamics in the small intestine of the young chick. British Poultry Science 86: 53-61.

Gous, R. M. and T. R. Morris, 2005. Nutritional intervention in alleviating the effects of high temperatures in broiler production. World's Poultry Science Journal 61:463-475. No. 3.

Hutagalung, R. T., 2000. The monetary crises and its impact on the development of poultry industry in Indonesia. Proceedings of the Australian Poultry Science Symposium 12. 74-81.

Johnson, R. J., 1987. Metabolisable energy: recent research. In: Recent advances in animal nutrition in Australia 1987, Armidale NSW, (Farrell, D. J. Ed.). Pp. 228-243.

Leeson, S., Summers, J. D. and J. J. Caston, 1992. Response of broilers to feed restriction or diet dilution to the finishing period. Poultry Science 71:2056-2064.

Morris, T. R., 2004. Nutrition of chicks and layers. World's Poultry Science Journal 60:5-11 No. 1.

Noy, Y. and D. Sklan, 1999. Energy utilization in newly hatched chicks. Poultry Science 78:1750 - 6.

Noy, Y., Geyra, A. and D. Skaln, 2001. The effect of early feeding on growth and small intestine development of the post-hatch poultry. Poultry Science 80: 912-9.

NRC (National Research Council). 1994. Nutrient requirement of poultry, 9th edition. National Academy of science, National Academy Press Washington, D. C.

Obioha, F. C., 1992. A guide to poultry production in the tropics. Acena Ventures Limited, Enugu. Pp 1-212.

Smith, A. J., 1990. Poultry. MACMILLAN PRESS LTD. London and Basingstoke. Pp 1-218.

Smith, A. J., 2001. Poultry (Revised edition). MACMILLAN EDUCATION LTD. London and Oxford. Pp 1-242.

Udedibie, A. B. I., 2003. In search of food: FUTO and the Nutritional challenge of Canavalia seeds. Inaugural lecture series 6. Delivered at the Federal University of Technology, Owerri on September 18th, FUTO Press Ltd. P. 7.

Uni, Z. and R. P. Ferket, 2004. Methods of early nutrition and their potential. World's Poultry Science Journal 60. 101 -111. No.

Studies in sub-Saharan Africa

Herausgegeben von Support Africa International Inc.
vertreten durch Baldur Pfeiffer und Franz-Theo Gottwald

Band 1   Uche C. Amalu / Franz-Theo Gottwald (eds.): Studies of Sustainable Agriculture and Animal Science in sub-Saharan Africa. 2004.

Band 2   Franz-Theo Gottwald / Susan Keino / Timothy Rotimi Fayeye (eds.): Fostering Subsistence Agriculture, Food Supplies and Health in Sub-Saharan Africa. 2007.

Band 3   George Ouma / Franz-Theo Gottwald / Isabel Boergen (eds.): Agrarian Science for Sustainable Resource Management in Sub-Saharan Africa. 2009.

www.peterlang.de

Franz-Theo Gottwald / Susan Keino /
Timothy Rotimi Fayeye (eds.)

Fostering Subsistence Agriculture, Food Supplies and Health in Sub-Saharan Africa

Frankfurt am Main, Berlin, Bern, Bruxelles, New York, Oxford, Wien, 2007.
199 pp., num. tab. and graph.
Support Africa International. Edited by Support Africa International Inc. Vol. 2
ISBN 978-3-631-57380-8 · pb. € 41.10*

The need in increasing food production to meet the food and nutritional demands of the ever growing population has necessitated this attempt to unveil the strategies of fostering subsistence agriculture in sub-Saharan Africa. To achieve this, the concept of subsistence agriculture, its place in a national economy and impact on poverty and health prevention, the various models and strategies were examined. Its intensity and dimensions are revealed in the role it plays not only in providing food with nutritional value, immediately available to the rural people, but also in creating the basis and formation stages for commercial agriculture. The farmers, governments, extensions, and other instruments involved in subsistence agriculture are the focal point of change

Contents: Strategies for subsistence agriculture under cultural and economic consideration · Agro-ecological approaches · Use of indigenous knowledge · Governmental farmer-oriented policies · Ecologically sustainable agriculture · Impact on health · Nutritional quality and food variety

Frankfurt am Main · Berlin · Bern · Bruxelles · New York · Oxford · Wien
Distribution: Verlag Peter Lang AG
Moosstr. 1, CH-2542 Pieterlen
Telefax 00 41 (0) 32 / 376 17 27

*The €-price includes German tax rate
Prices are subject to change without notice
Homepage http://www.peterlang.de

Peter Lang · Internationaler Verlag der Wissenschaften